CASS SERIES: STUDIES IN INTELLIGENCE
(Series Editors: Christopher Andrew and Michael I. Handel)

DIEPPE REVISITED
A Documentary Investigation

Also in this series

Codebreaker in the Far East
by Alan Stripp

War, Strategy and Intelligence
by Michael I. Handel

A Don At War (revised edition)
by Sir David Hunt

Controlling Intelligence
edited by Glenn P. Hastedt

Security and Intelligence in a Changing World: New Perspectives for the 1990s
edited by A. Stuart Farson, David Stafford and Wesley K. Wark

Spy Fiction, Spy Films and Real Intelligence
edited by Wesley K. Wark

From Information to Intrigue: Studies in Secret Service Based on the Swedish Experience 1939–45
by C.G. McKay

The Australian Security Intelligence Organization: An Unofficial History
by Frank Cain

Intelligence and Strategy in the Second World War
Edited by Michael I. Handel

Policing Politics: Security Intelligence and the Liberal Democratic State
by Peter Gill

DIEPPE REVISITED
A Documentary Investigation

JOHN P. CAMPBELL

LONDON AND NEW YORK

First published 1993 by FRANK CASS & CO. LTD.

Published 2019 by Routledge
2 Park Square, Milton Park, Abingdon, Oxon OX14 4RN
52 Vanderbilt Avenue, New York, NY 10017

Routledge is an imprint of the Taylor & Francis Group, an informa business

Copyright © 1993 John P. Campbell

All rights reserved. No part of this book may be reprinted or reproduced or utilised in any form or by any electronic, mechanical, or other means, now known or hereafter invented, including photocopying and recording, or in any information storage or retrieval system, without permission in writing from the publishers.

Notice:
Product or corporate names may be trademarks or registered trademarks, and are used only for identification and explanation without intent to infringe.

British Library Cataloguing in Publication Data

Campbell, John P.
 Dieppe Revisited: Documentary
 Investigation. – (Studies in
 Intelligence Series)
 I. Title II. Series
 940.54

Library of Congress Cataloging-in-Publication Data

Campbell, John P.
 Dieppe revisited : a documentary investigation / John P. Campbell.
  p.   cm. -- (Cass series--studies in intelligence)
 Includes index.
 ISBN 0-7146-3496-4
 1. Dieppe Raid, 1942. 2. Dieppe Raid, 1942--Sources.
 3. Deception (Military science) 4. Military intelligence--Great Britain. I. Title. II. Series.
 D756.5.D5C36 1993
 940.54'21425--dc20   92-19265
                                                   CIP

ISBN 13: 978-0-7146-3496-8 (hbk)

Contents

Acknowledgements	vi
List of Maps	viii
A Note on Naval Terminology	ix
Abbreviations	xi
Introduction	xiii
1 Another Book about Dieppe?	1
2 Spy Stuff – Agents, Cover and Deception	41
3 Channel Watch 1942	78
4 Radar and the Raid	122
5 Most Secret and Other Sources	159
6 Dieppe and D-Day	196
Glossary of Codenames for Operations	234
Index	235

Acknowledgements

I should like to thank the Trustees of the Liddell Hart Centre for Military Archives for permission to quote from the Liddell Hart Papers; the Trustees of the Broadlands Archives Trust and Lord Tweedsmuir for permission to include quotations from the Mountbatten Papers; and the Trustees of the National Maritime Museum, Greenwich, for permission to publish material from the Papers of Admiral T. Baillie-Grohman. I am also grateful to the Controller of Her Majesty's Stationery Office for permission to reproduce three of the 'Plans' contained in the Naval Staff History (Battle Summary No 33) of the raid on Dieppe.

My main debt of gratitude in a work of this description is bound to be to the professional staffs of the four main repositories where I did my research: the Public Record Office in London, the National Archives in Washington, the National Archives of Canada in Ottawa, and the Bundesarchiv-Militärarchiv in Freiburg-im-Breisgau. Time and again I was reminded that the archivist is the historian's best friend. Harry Rilley in Washington was particularly helpful in tackling the German naval documents on microfilm. Although my emphasis was on official records, there was always the temptation to appeal for elucidation of some particularly thorny point to a knowledgeable authority, a temptation to which I yielded on a number of occasions. Patrick Beesly, Sir Harry Hinsley, Sir Michael Howard, Ewen Montagu, Captain S. Roskill and Willi Weber were all generous with their replies to my queries, some of which probably struck them as rather far-fetched. I had the honour not only of corresponding with but actually meeting Professor R.V. Jones and David Mure. The only veteran of the Dieppe raid I interviewed

ACKNOWLEDGEMENTS

was Jack Nissen, who patiently explained some of the basic facts of electronics to this scientific ignoramus; it is accordingly with genuine regret that I find my reading of the documents at odds with Jack's recent claims based on his exploits on 19 August 1942. Another expert who dazzled me with science was J.R. Robinson, who was good enough to read my chapter on radar.

Finally, three fellow historians made a contribution to the writing of this book: Brian Villa kept to the informal agreement we made in Vancouver in 1983 to 'divide' the topic and unintentionally acted as a goad by publishing his own book to great acclaim in 1989; and Bob Johnston and Dick Rempel, the best of friends and colleagues, read the text and offered invaluable advice on everything from umlauts to organization. My deepest debt of all is to my wife, Jennie, who shared the joys and rigours of many a research trip, and who never once complained that it was taking so long to finish this project.

John P. Campbell,
Ancaster, Ontario

List of Maps

frontispiece The English Channel	xvi
Map 1 Operation Jubilee, 19 August 1942	99
Map 2 Action with Enemy Forces	144
Map 3 Plot of Enemy Convoy	168

A Note on Naval Terminology

Much confusion can be averted by bearing in mind that German T(orpedo)-boats were really small destroyers: in fact the earliest Type (1923), to which *Seeadler* belonged, was initially classified as such. T-boats were slightly smaller in displacement than the Hunt Class of escort destroyers which began to appear in the Channel in 1940/41. *Calpe*, for instance, was a Type II Hunt displacing 1050 t standard, with a complement of 168; *Iltis*, 924 t standard, with a ship's company of 127. The T-boats were capable of speeds in excess of 30 knots, compared with 27 knots for *Calpe* and her Class, which had no torpedo armament. The true German equivalent of the British Motor Torpedo Boat was the formidable E-boat, or S(chnell)-Boot, with two fixed torpedo tubes forward, a speed in excess of 35 knots, and a crew of about 20, depending on Type. Altogether the Kriegsmarine commissioned 249 E-boats. With a length of 114 ft, they were easily confused in PRU photographs of dockyards with the much slower R(äum)-Boote of 116 ft. R-boats were small motor minesweepers with a complement of 34/38, no torpedo armament but a heavier gun armament than the E-boats. Like the even slower V(orposten)-Boote, which were mostly armed trawlers, they were often pressed into service as convoy escorts. M-Class minesweepers were excellent, versatile ships of 680 t standard and a complement of 95/113, capable of being used as escorts, submarine chasers, or minelayers. U-Jäger did not denote a standard A/S type of vessel but covered all that were equipped for A/S duties, such as UJ 1404, a converted deep-sea trawler equipped with 60 depth charges, detection gear and an 88 mm gun as main armament. Sperrbrecher were usually

A NOTE ON NAVAL TERMINOLOGY

converted tramp steamers whose crews had the unenviable task of detonating mines by contact. British MTBs and MGBs were powered by noisy petrol engines. It was in an attempt to overcome this disadvantage that the Steam Gunboat was designed (34 knots, 175 t standard, crew of 27), but the SGB proved to be rather too vulnerable to be called a complete success.

Abbreviations

Where official documents are concerned, the overwhelming majority of those consulted were conveniently concentrated at the Public Record Office (PRO) at Kew and the National Archives (NA) in Washington. ADM, AIR, CAB, DEFE, FO and WO are all PRO classes. German naval (T 1022), diplomatic (T 120), and military (T 78, 315 etc) records are readily available in the Microfilm Room at the NA. This particular pattern of research conformed to nothing more rational than mere personal convenience: many of the German documents must now be available in London at the PRO, Imperial War Museum or elsewhere, and obviously the NA (RG 331) copy of the Rattle file is not the only surviving one. The blessings for the researcher derived from the fact that the war on the Allied side was largely run by Joint as well as Combined staffs and committees are legion. Not all German naval records, by the way, are included in the T 1022 series, the most important exception in this case being the War Diary of Befehlshaber der Sicherung West, the original of which (M 444) has been preserved at the Bundesarchiv-Militärarchiv in Freiburg. At the National Archives of Canada, Army war diaries (RG 24) proved to be particularly valuable.

The following are among the abbreviations used in the text and endnotes:

ADI(Sc)   Assistant Director of Intelligence (Science), Air Ministry
AOK   Armeeoberkommando [Army HQ]
Ast   Abwehrstelle [Abwehr station in Germany or Occupied Europe]

ABBREVIATIONS

B-Dienst	Funkbeobachtungs-Dienst [Naval Wireless Intelligence Branch]
BSW	Befehlshaber der Sicherung West [Naval Commander, Western Defences]
C	Head of British Secret or Special Intelligence Service (MI 6)
Chefs	Chefsache [Top Secret]
COS	British Chiefs of Staff
COSSAC	Chief of Staff to Supreme Commander (Designate)
DNI	Director of Naval Intelligence
FdS	Führer der S-Boote [Flag Officer, MTBs]
FS	Fernschreiben [Teleprint]
GC & CS	Government Code and Cypher School
GenStdH	Generalstab des Heeres [Army General Staff]
gKdos	geheime Kommandosache [Secret]
ID	Infanterie-Division
IR	Infanterie-Regiment
ISSB	Inter-Service Security Board
JG	Jagdgeschwader [Fighter unit of 3 groups]
JIC	Joint Intelligence Sub-Committee (of COS)
KG	Kampfgeschwader [Bomber unit of 3 groups]
KTB	Kriegstagebuch [War Diary]
MarGpW	Marinegruppenkommando West [Naval Group West]
MSS	Most Secret Source (Ultra)
Ob	Oberbefehlshaber [C-in-C]
ObdM	Oberbefehlshaber der Kriegsmarine [C-in-C Navy]
OIC	Operational Intelligence Centre (Admiralty)
OKH	Oberkommando des Heeres [Army High Command]
OKW	Oberkommando der Wehrmacht [Armed Forces High Command]
Ozet	Ortungs-Zentrale [Radar Plotting Centre]
PR	Photographic Reconnaissance
Pz	Panzer [Armoured]
RDF	Radio Direction Finding
R/T	Radio Telephony
SichDiv	Sicherungs-Division [Naval Defence Division]
Skl	Seekriegsleitung [Naval Staff]
V-Mann	Vertrauens-Mann [Secret Agent]
W/T	Wireless Telegraphy
Y	Wireless Intelligence

Introduction

Like Dunkirk, Dieppe is one of those towns and cities touched by the Second World War whose names have had a different resonance ever since. This most agreeable of the French Channel ports was known before 1939 for its sunsets and the 'white magic' of its light. One outstanding artist after another was drawn to Dieppe – Delacroix, Turner, Bonington, Monet, Degas, Pissarro, Renoir, Whistler, Gauguin, Sickert. Among the writers who relished its understated charm were Chateaubriand, Flaubert, Turgenev, Oscar Wilde, Proust and Virginia Woolf. Saint-Saëns was born there, Liszt and Rossini were visitors.[1] Sadly enough, this rich cultural tradition and all the varied delights of harbour, cliffs and beach have been overshadowed since 1942 by the violent events of a summer's morning, while at the same time the ties binding Canada to this part of Normandy for almost 300 years have taken on a new poignancy.

It has been claimed that more has been published about the raid in 1942 than about the invasion in 1944. Whatever the truth of it, this is still an astonishing observation. No mere raid, after all, could conceivably compete with the Normandy invasion – or for that matter the Dunkirk evacuation – in strategic significance. What excuse, therefore, could there be for making a further contribution to this historiographical imbalance? There is no facile explanation, such as the approach of the raid's fiftieth anniversary or outrage at some supposed victimization of Canadians. Rather, this contribution to the Dieppe canon

---

[1] See John Russell, 'Homage to the Discreet Charm of Dieppe', *New York Times*, 24 May 1992.

INTRODUCTION

can best be described as an example of the indirect approach to a topic, being the outgrowth of an interest that began with general aspects of the history of the war rather than with the raid itself. Only in the 1970s and 1980s, it should be remembered, did serious research in official sources become possible for non-official historians; and only in 1972, coincidentally, was the veil grudgingly lifted for the first time from the history of wartime intelligence in its numerous manifestations. Not only was there a good deal more to be said, it quickly became apparent, about campaigns, operations and military reputations, but the opportunity to do so had unexpectedly been opened up. *Dieppe Revisited* began with the premiss that it would be a useful scholarly exercise to examine some inherently interesting subjects – intelligence and deception, radar, naval and air operations – in the context of the English Channel in 1942, with an emphasis on documentary rather than published sources and always with equal attention to German files and war diaries.

The Dieppe raid was simply the obvious peg on which to hang such an investigation. There was always the risk, naturally, of finding nothing new, or even nothing at all, to say, not to mention the certainty that sooner or later the documents would run out, whether for reason of loss or destruction during the war or continued classification fifty years after its conclusion. As it turned out, there was plenty to write about with the material available, despite the earlier attentions of at least three official historians. Yet *Dieppe Revisited* will disappoint those who enjoy their history with a revisionist twist, for it is unapologetically confirmatory in tone and detailed in focus. Jubilee remains a badly planned operation which was launched largely out of sheer determination to lay on a sizeable raid on the part of the Chief of Combined Operations and his immediate staff and, it might be added, of over-confidence on the part of the military. Although its security could charitably be described as chequered at best, the timing and location of the raid took the enemy by surprise, notwithstanding the wearisome efforts of some historians to prove otherwise. The lessons it yielded for D-Day, always a flash-point of controversy, are still difficult to define once allowance is made for those that should never have had to be relearned at such cost and others that would surely have been learned anyway well before 1944. Still, Jubilee

INTRODUCTION

could not but have contributed to the planning of the assault on 6 June 1944.

In short, this book has no claim to be a comprehensive or even a balanced one about the raid on Dieppe. There are several of those in print already. It is more of a self-indulgent examination of some hitherto obscure or neglected aspects of what for want of a better description has been termed the raid's operational context; research was largely guided by the accessibility of documentary sources; and exclusive attention to matters of immediate relevance to Jubilee was not meant to be one of its characteristics. But if the reader finds the final result of this approach half as enjoyable as the researcher did the process of working towards it, then the book will have served its main purpose.

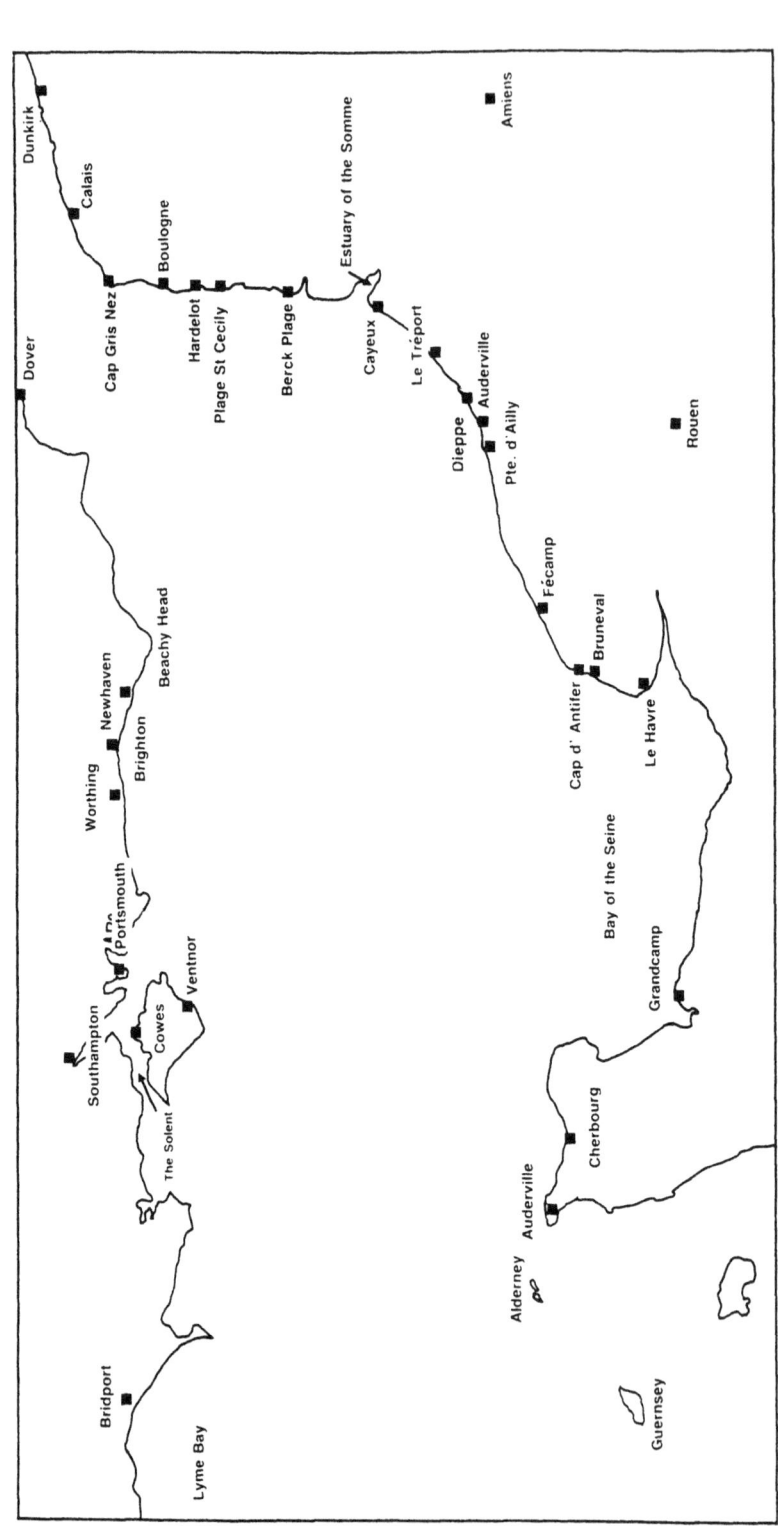

The English Channel

1

Another Book about Dieppe?

Arguably, one of the last things the history of the Second World War needs is yet another book about the raid on Dieppe on 19 August 1942. Canada's single greatest tragedy of the war has always claimed its share of the melancholy fascination surrounding military disasters. Even at the time, Operation Jubilee inspired observations suggesting a greater significance than the size of the forces engaged would normally have warranted. Hitler showed quite a shrewd sense of military history by comparing Dieppe with the first tank battle at Cambrai in November 1917; in both cases, he claimed, defeat was the result of inflexibility in command and failure to plan beyond the first few hours, rather than of any technical shortcomings in the new equipment used. Air Vice-Marshal Trafford Leigh-Mallory, Air Force Commander for Jubilee, showing rather a less impressive grasp of the history of the two World Wars – and no sense whatsoever of bathos – congratulated his military counterpart, Major-General J.H. Roberts, on 'the magnificent feat of arms' in bringing off 'a Gallipoli and a Dunkirk all in the same day'. Another Major-General, P.C.S Hobart, wrote to Liddell Hart on 25 August 1942:

> What I don't understand is why Dieppe? A raid is either 'to obtain information' and destroy some worthwhile objective or it is to train one's own troops. In the latter case one would not select a strongly defended sector. In the former, what was the objective? Evidently we did not reach it. It all sounds pretty Passchendaele to me.

The chief reason for the spell still cast by Jubilee is of course the tragic casualty rate suffered by the Allied forces involved, particularly the Canadian Army. Just over 900 Canadians were killed and more than twice that number (1,946) captured out of a force of roughly 5,000, not all of whom landed.[1] Raids, too have always attracted disproportionate interest by reason of their dramatic and self-contained nature. Add to that the fact that this particular raid might almost have been deliberately staged to encourage the propagation of rumour and legend and there is no need to wonder that it has never suffered from neglect on the part of historians, journalists, mythologists and the producers of documentaries for TV.

The history of the origins and planning of the raid and of the events of 19 August is well trodden ground, having already been covered by Colonel C.P. Stacey, Jacques Mordal, Terence Robertson, Ronald Atkin, Norman Franks and others.[2] Stacey's Official History of the Canadian Army got the subject off on a sound scholarly footing in 1955, although he did not succeed in defusing its more controversial aspects. Canadians have always laid claim to Dieppe as a distinctively national calamity – understandably so. 'Its [the raid's] interpretation needs an innate understanding of the Canadian psyche', one of Robertson's reviewers pontificated; Robertson, he added, had not lived in Canada long enough to absorb the ethos of 'an affair that belongs to the warp and woof of Canadian society more than to the grand movement of the war'. Dieppe also remains a sore point for the biographers of its major participants, as the contrasting treatment accorded it by Nigel Hamilton in *Monty* (1982) and Philip Ziegler in *Mountbatten* (1985) amply demonstrates. In keeping with the historiography of its subject, Brian Loring Villa's penetrating study of Mountbatten's role in the launching of Jubilee (1989) led to an often acrimonious exchange of correspondence with Ziegler and others.[3] *Dieppe Revisited* is not another narrative history of the raid; nor does the book attempt to fix responsibility for the losses or aspire to define the place they occupy in the minds and souls of Canadians. Instead, it is the outcome of what began with curiosity about the operation's context and an interest in the evolutionary history of the release of official records for the Second World War.

There is nothing conceptually grandiose about the matter of operational context. What is meant is simply how the war worked, so to speak, a level or two below the political and strategic. In the case of 2 Canadian Infantry division, 19 August 1942 was preceded by months of training and anti-invasion exercises and followed by the long process of restoring battleworthiness. Across the Channel, exercises and fatigues continued to be the main preoccupations of Lieutenant-General Konrad Haase's 302 Infanterie-Division (ID) until its departure for Russia at the end of the year. From a strictly military point of view, Jubilee stood out as an isolated operation halfway between Dunkirk and D-Day. This was obviously not the case, however, for the naval and air forces on both sides of the Channel, for whom it was a dramatic interlude in the course of prolonged campaigns in which almost all the action at sea took place at night and most of the air activity by day. What makes these lopsided campaigns interesting is not their scale or strategic significance so much as their complexity. The air between France and England was constantly humming with all manner of communications and radar signals, making the Channel a very intimate theatre of operations in which both the Germans and the British tended to deploy their latest equipment along the coast. Moreover, during 1942 there was still a certain balance in the sense that the Allies had not yet gained the upper hand to the extent that they did later in the war. Nineteen forty-two was the only year of the war when there were invasion craft in both the English and French Channel ports, the only year other than 1944 when Hitler took the threat of a cross-Channel invasion at all seriously. It was also pre-eminently the year of the cross-Channel raid, planned as well as executed. Whereas the history of the war has been exhaustively covered so far as its major or dramatic operations are concerned, the same is not true of many of the unspectacular and sometimes marginal activities and events that made up its texture in any theatre – offensive and defensive mine-laying, mine-sweeping, aerial reconnaissance, radar development, the planning of operations that were never launched – some of which might yet reward the kind of detailed examination that is now possible. Jubilee is a good selection for taking an eclectic look at the machinery of the war in the West, if only because it involved all three services on both sides and took place amidst the strategic uncertainty of 1942.

Before identifying those features of its context that have been overlooked or underplayed and before raising the question of the survival and accessibility of documentary sources, it is necessary to review, however sketchily, something of the operation itself. The most amazing thing about the raid was that it ever came off at all, since it did so only after the most extraordinary series of fits and starts. First planned as Operation Rutter and rehearsed in the Yukon I and II exercises in June, it was postponed and then modified from a two-tide to a one-tide operation on 5 July, only to be cancelled because of the weather forecast two days later; after being resurrected at a highly secret meeting on 11 July, it was replanned in some respects and finally launched as Jubilee. The possibility of remounting the operation in the event of such a cancellation had, incidentally, been brought up by Lord Louis Mountbatten, Chief of Combined Operations (CCO), on more than one occasion before 7 July. Indeed the ultimate origin of Jubilee might be traced to a suggestion which Mountbatten agreed to in principle on 23 January 1942. A *repeat* raid on Dieppe was to be undertaken in July in the hope of inflicting casualties on the engineers and specialists engaged in repairing the damage from an earlier raid in June.[4]

The initial postponement of Rutter on 15 June following the unsatisfactory Yukon I called for some quick footwork by the security authorities. The Inter-Service Security Board (ISSB) recommended that the troops detailed for Rutter should be led to believe that there had been no postponement and that they were continuing training for a cross-Channel invasion or Second Front; other exercises, including another full-scale one, should be held with the additional objective of confusing the enemy. Still, postponing an operation is always tricky, especially if some units earmarked for it have already been briefed. On 22 June, Mountbatten was invited by his fellow Chiefs of Staff (COS) to report 'as a matter of urgency' on the extent to which Rutter's security might have been compromised by the requirement to brief airborne officers and aircrew well in advance. The COS decided as a result of CCO's investigation that there was no need for Rutter to be cancelled because of any possible leakage but that pilots and observers in the two Wellington squadrons who had been briefed should not take part in raids over Germany until after the operation.[5]

Generally speaking, security on the south coast of England left a lot to be desired, in the sense that a good deal could seemingly be deduced simply by shrewd observation and careful attention to careless talk. A group of thirty 'ordinary citizens' of a 'Home Guard Guerrilla Platoon' set out to find out how much they could discover in a week by engaging servicemen, lorry-drivers and others in casual conversation; their findings were published in the *Daily Mail* on 4 July, no doubt to the embarrassment of the authorities. Commando training was under way on 'a certain island', which could only mean the Isle of Wight, 'radio location stations' had been set up at X and Y, troops were ready to embark for France, and so on. Mountbatten's Military Adviser pointed out that any spectator on the porch of the Gloucester Hotel in Cowes on 2 July would have had no difficulty determining that an operation was about to take place when confronted by the sight of troops embarking on the Landing Ships Infantry (LSI) in the harbour, followed by officers strapping on side-arms. The locals had plenty to say about the erection of dummy funnels and other camouflage on the LSIs in full view of shore. Even the clerk who had typed the Rutter Operation Orders, the ISSB later discovered, had disappeared until 1 August.[6]

Worse still, of course, 'several hundred' people, according to the Combined Report on Jubilee, had known for just over forty days between Rutter and Jubilee that a raid on Dieppe had been in the offing. The true total must have been more like several thousand. Rumours flew thicker than ever, so that by the start of August security officials had difficulty distinguishing between those that might be current and those that had originated at the time of Rutter and were still going the rounds. Canadian Army Field Security was deluged with reports of careless talk immediately after the BBC broadcast announcing the raid at 0800 on 19 August. But the naval officer in charge (NOIC) at Newhaven was so alarmed by the speculation on Saturday, 15 August, about an impending raid on the French coast that he could only hope that 'the news being spread about could not pass over to the enemy in time to affect the operation'.[7] Ironically enough, there was no lack of security consciousness, at least at the top; the Vice-Chief of Combined Operations somewhat theatrically locked his door and disconnected his phone before giving Lord Lovat a personal briefing at Combined Operations HQ (COHQ) in Richmond Terrace just off Whitehall. Evidently there was to be no repetition of an earlier incident, when journalists arrived at Dover asking for news of 'last night's raid on Boulogne' two days *before* Operation Abercrombie on 21/22 April.[8]

The War Office had commissioned a film to dramatize the message that 'walls have ears', in which a German spy, played by Mervyn Johns, got wind of a plan for a Commando raid on the French coast, with consequences not unlike those on 19 August. *Next of Kin*, directed by Thorold Dickinson, was originally intended to be shown only to members of HM Forces, and for many of the Canadian troops taking part in Jubilee it later came to symbolize the truth about the raid. 'Comment on Dieppe – see *Next of Kin*', one of them wrote; '. . . the Germans were there waiting for us at ten past five in the morning, with machine-guns.' So emphatic were the letters to Canada on this point that they convinced the military censors that information about the raid had indeed reached the enemy beforehand. Churchill, according to one source, had qualms about an incident in the film that might be taken to have foreshadowed Jubilee and had it subjected to a military jury before it was approved for general release. *Next of Kin* actually seems to have been fairly widely distributed before Jubilee, judging by the fact that the thirty 'ordinary citizens' who conducted their own counter-intelligence survey claimed it as their inspiration.[9] At any rate, the Prime Minister was correct in anticipating that Jubilee would quickly acquire a reputation at least partly based on a fancied resemblance between dramatic art and reality. There were plenty of comments just after the event like Robert Bruce Lockhart's, to the effect that the Germans had known about the operation except perhaps for the actual date. For months MI 5 and the ISSB were kept busy investigating possible leaks and indiscretions. 'Responsible authorities in the UK', wrote Stacey in 1946 (without identifying them), 'had thought it decidedly probable that some information might have reached the Germans concerning Operation Rutter.' Before the end of March 1943, there had already been two major investigations into Jubilee's security at the request of the Joint Intelligence Committee (JIC).[10]

At the tactical level, the planners of Jubilee could never be accused of lack of attention to detail. The Combined Plan (JJ One, 31 July) was at the centre of the three detailed plans which in turn were the basis for Operation Orders issued by the

Force Commanders to their respective commands. The 'Detailed Military Plan' (JG One) had twenty appendices, one of which – Appendix C, 'Allotment of Personnel, Equipment and Stores' – alone ran to fifty pages. The naval 'Outline of Operation' (JNO 1) also went into much greater detail than was normally the case, chiefly because of the complexity of the operation and the lack of training and experience among the landing-craft crews, threatening a breakdown in communications through signals congestion. There was no question that the naval side of Jubilee was a difficult enough undertaking even without allowing for active enemy interference. Some of the landing craft, such as the 23 R-craft (LCP(L)) in 5 Group carrying No 3 Commando to the easternmost beach at Berneval, were to make their own way across the Channel, but most of the infantry would be carried in nine LSIs, mostly converted cross-Channel steamers carrying between 200 and 300 troops and six to eight assault craft (LCA) on davits. The LSIs, moreover, would reach their lowering points some ten miles off Dieppe only after following two flotillas of minesweepers through a suspected minefield. In company with them would be 24 Tank Landing Craft (LCT) on their first operation, plus an assortment of small craft in dispersed groups, all moving at their own speeds. The covering force of eight small Hunt-class destroyers could charitably be described as modest. To cap it off, every sizeable cross-Channel operation before Jubilee had run into difficulty finding the right beach, let alone co-ordinating the movements of such a large and disparate force.[11]

The pages of Embarkation Tables, Departure Timetables, Withdrawal Timetables (for both the one- and two-tide versions of the raid) could not obscure the essential truth about the Jubilee assault: the four flank landings, two on either side of the town, depended totally on surprise. Success was a matter of landing in the dark at 0450 (British Summer Time)[1] without the enemy's being aware of the approach to the beaches – or at worst, while it was still too dark for those manning the defences to bring their weapons to bear accurately. The two outer landings

---

[1] All times in this text are BST unless contained within a direct quotation from German sources; BST was one hour ahead of Greenwich Mean Time and one hour behind Deutsche Sommerzeit.

at Berneval and Varengeville by 3 and 4 Commando respectively were intended to neutralize coastal batteries, an assignment they took over from Rutter's parachute troops, while the inner flank landings at Puits and Pourville were directed against the defences on the two headlands dominating the sea approaches to the town, its harbour and its beach. The main landings on the town beach (divided into Red and White) would take place half an hour after the flank landings, by which time it would certainly be light enough to bring weapons to bear accurately; one reason given for this interval was the lack of sea-room for all the landing craft to form up for a simultaneous touchdown at 0450 on all five beaches. In short, the assault straddled the first glimmering of daylight, coming after Nautical Twilight (0431) and well before Sunrise (0550), and presupposed standards of synchronization and navigation that a much smaller, specialist force would have been hard-pressed to meet.

The main landings have always been categorized as 'frontal', although all amphibious assaults could be so described in the sense that they offer no opportunity to fall back to regroup or manoeuvre to outflank the beach defences. Some contested landings, however, are decidedly more 'frontal' than others because of the strength of the defences on shore. The main assault at Dieppe certainly fell into the latter category, in addition to suffering from a lack of the fire support that would normally have compensated for loss of surprise. At a planning meeting on 5 June, the preliminary raid by heavy bombers that had been included in the Rutter Appreciation and Outline Plan approved by the COS on 13 May was called off; whatever advantages the bombing offered, it was decided, would be outweighed by the disadvantages of alerting the defences and obstructing the streets with rubble – an opinion the three Force Commanders continued to hold after the operation. The Admiralty had already flatly refused to risk a replay of the sinking of the *Prince of Wales* and *Repulse* by committing a battleship or cruiser to an operation in the confined waters of the Channel. The Naval Force Commander for Rutter, Rear-Admiral H.T. Baillie-Grohman, asked Mountbatten for a battleship but was told in reply on 4 June that it was above all important to be able to stage a victory untarnished by a major disaster – an explanation Baillie-Grohman found 'pretty feeble'.[12] Consequently, fire support was reduced to two passes by five squadrons of cannon-firing Hurricanes that lasted all of ten minutes and a 'bombardment' by a monitor and the 4-inch guns of the destroyers backing up the landing craft. Even at that, too much gunfire by the two HQ destroyers, *Calpe* and *Fernie*, was thought 'undesirable' as it was liable to interfere with the numerous W/T sets specially installed for the occasion; and since the time of arrival of the Hurricanes was fixed, the troops could not afford to arrive early or more than fifteen minutes late if they were to enjoy such support as these offered.

One of a number of ironical features of Jubilee was that its naval and air aspects inspired more preliminary misgivings than its military. Alastair Buchan, who was involved as a member of the Canadian Army staff planners and took part in the raid, wrote to Mountbatten after watching the BBC-TV programme on Dieppe in August 1972:

> I can remember the illusions that were entertained by our staff as well as yours about the ease with which beach and port defences could be overcome. I don't see how we could ever have overcome those illusions without putting them to the test.[13]

This brief comment contains much of the truth about Dieppe. Leigh-Mallory, by contrast, was worried by a number of aspects of the operation, including the fact that an assault and withdrawal on the same day would submit his squadrons to a greater strain than would a more prolonged operation. Hughes-Hallett also made no bones about anticipating a hazardous operation, which helps to explain why his instant reaction was to be both relieved and pleased by the precision achieved in the landing on Red and White beach, the like of which he had never witnessed on exercises. The penalty, alas, for not quite achieving the necessary standard of precision was to be tragic. The most experimental feature of the entire operation was the landing of the Churchill tanks of 14 Canadian Army Tank Battalion on the same beach and on the heels of the first wave of infantry, and it turned out to be one of the least successful. The infantry may have been a minute or two late in touching down after their two-hour approach from the lowering point, but the first three LCTs were up to fifteen minutes late. The first wave of nine tanks were not there to provide the essential continuity of fire to engage pillboxes and prevent the Germans from manning their positions. Of the 24 LCTs, ten disembarked 29 tanks, but only about half of these managed to cross the sea-wall to the promenade and none reached the streets of the town. Heavy casualties among the sappers landing from the LCTs prevented the blasting of obstacles. Some tanks lost adhesion in the shingle but most of the damage to their tracks was apparently done by the accurate fire of anti-tank weapons. None were taken off again and the use of the demolition charges with which the tanks were fitted was apparently inhibited by the fact that many were being used to shelter wounded. Not the least questionable feature of the raid was the commitment of a new type of tank in circumstances that almost guaranteed that some would be captured intact. For weeks afterwards a suspicious concentration of armour near Yvetot showed up in RAF reconnaissance photographs until an 'acute source' on the spot identified the impression of Churchill tracks. Even before then, Marks I, II and III had been the subject of a scornful technical evaluation by German specialists.[14]

But the decisive delay, producing a ripple effect for the entire force, came earlier at Puits (Blue beach), about two miles east of the harbour entrance. Hughes-Hallett considered this landing to be so vital that he recommended on 16 August that the operation should be called off if the landing force heading for Blue beach were lost on passage – an appreciation, incidentally, not shared by the military staff.[15] The east headland (Bismarck) commanded the harbour entrance and the Germans would be able to bring down enfilade fire on Red and White beach unless their battery and machine-gun positions were neutralized or captured. In the plan for Rutter, the parachute drops were timed for 0330; more to the point, the inner flank landings were to take place at 0415, an hour before the main landing, which would have given the troops on the flanks at least a fighting chance to carry out their tasks. In fact the Naval Operation Orders stated quite clearly that troops from Green and Blue beaches should have reached the batteries on the two headlands by the time of the main assault at 0515. There was no such assurance in the case of Jubilee, and it was surely unrealistic to expect the Royal Regiment of Canada landing at Puits to suppress, never mind capture, the 4-gun battery (Rommel), a heavy anti-aircraft battery, and light anti-aircraft and machine-gun positions in a mere half-hour. The Operation Orders more or less conceded this point. The orders for Phase II of prearranged naval fire support in Sector V – positions on or in the cliffs eastwards of the harbour – called for *Garth*, one of the Hunts, and the monitor *Locust* to continue fire for fifteen minutes *after* touchdown on the main beach; there was, of course, to be no fire support for the approach and landing at Blue beach, since Phase I of the bombardment applied only to Red and White beach from 0510 to 0520. In short, even if everything had gone according to plan, it was accepted that there was every chance that the positions on Bismarck would not have been dealt with decisively when the initial infantry assault and first LCTs touched down on Red and White beach. Mortars were supposed to open up with smoke and HE from Blue beach at 0505; but a 3-inch mortar detachment landing at 0520 on Red beach was also detailed to fire on the east headland.[16]

Little at all went according to plan. The landing craft from *Queen Emma* and *Princess Astrid* first formed up on the wrong motor gunboat (MGB), then speeded up to make up time (thereby increasing engine noise), and arrived at Blue beach twenty minutes late, without mortars and 135 men short. A landing meant to take place under cover of darkness had instead to be attempted in half-light against an alert defence. There was no back-up plan for the eventuality of a landing in daylight such as the No 4 Commando Operation Order contained. The small garrison holding Puits had been on a night exercise and were about to stand down when a spectacular fire-fight had broken out at sea at 0340, died down after a few minutes, and then flared up again at 0415. A coastal convoy from Boulogne preparing to enter Dieppe on the tide had collided with 5 Group, scattering the R-craft, badly damaging Steam Gunboat (SGB) 5, the group leader, and throwing off the timing of the landing at Berneval. Although the battalion commander was unaware that an amphibious operation was in progress, he kept his troops on the east headland and at Puits standing-to, with the result that the landing never had a chance, the LCAs being peppered with small-arms fire even as they closed the beach. The naval bombardment of the east headland was ineffectual. *Locust* was held up and did not appear on the scene until 0545, when she was taken under fire by the heavy AA battery and driven off, having sustained casualties. *Garth* was unable to fire in indirect support through smoke, not being equipped with a 'bearing plot', and was driven off when she closed to provide direct fire support to Blue beach. Meanwhile, the Forward Observation Officer in contact with *Garth* was helplessly pinned down in front of the wire and seawall, and sniper and machine-gun fire wiped out the 3-inch mortar crews once landed. It was soon clear that there could be no question of having *Locust* or any other vessel enter Dieppe harbour to destroy installations and remove invasion barges.[17]

Worse still, fire from Bismarck onto the main beach picked up in intensity as soon as the smoke laid by Boston and Blenheim aircraft from 0510 to 0540 began to clear. Dropping smoke-bombs from fifty feet proved to be effective but after the commitment of the four reserve aircraft circling Selsey Bill at 0520, there was no backup. A request relayed by *Calpe* at 0605 for more smoke on Bismarck could not be met until 0740 because the remaining smoke aircraft were either armed with Smoke Curtain Installation (SCI) to cover the withdrawal or were converting to SCI from bombs, a process that took over an hour on a Blenheim. The three Blenheims despatched at 0740 were actually recalled at 0817 while en route following a signal from the Military Force Commander, apparently under the impression that more headway had been made against the east headland than was the case. Not surprisingly, Leigh-Mallory later decided that he could have used three times the number of smoke aircraft available.[18]

In the other inner flank landing at Pourville (Green beach) surprise was achieved but the Germans recovered quickly enough to prevent penetration of the trench system at Quatre Vents farm on top of the west headland. Consequently Red and White beach was swept by fire from the headlands at either end and the troops never had a chance to cross the seawall in force or, indeed, to achieve any useful purpose at all. The dual-purpose AA guns on both heights were used to fire directly onto the beach below; mortar fire was particularly deadly, coming from covered positions, often on fixed lines that were well co-ordinated with machine guns. It was never possible to set up ambulance collecting points on the beach or to carry out anything more than 'casual first-aid'. The interval between being wounded and being operated on in hospital in England was anything from twenty to thirty-six hours.[19]

A point worth making, though, is that it was surely unrealistic to expect to approach any of the landing beaches at 'ten past five', as the 'Comment on Dieppe' correspondent evidently did, without drawing fire. Even if the flank landings had gone according to plan, tactical surprise for the main landing was not to be expected. Yet many veterans of the operation found it difficult to accept that an aroused garrison was not self-evident proof of a well-prepared ambush laid hours, possibly days, in advance. For instance, the lack of movement in the enemy's rear was interpreted to mean that he had anticipated the raid and reinforced Dieppe; his tactics were evidently to allow the troops to land and then open up with concentrated mortar and machine-gun fire. The Germans took a delight in reinforcing this impression for propaganda purposes; and individuals who came into contact with POWs followed the same line. 'Our Army knew you were coming and were well prepared', a doctor told a wounded Canadian in hospital in Rouen. 'Your security was very bad.'[20] Impressions formed in the heat of battle or shortly after being taken prisoner are not easily eradicated, so that from the moment the first units arrived back in England, any rumour or sinister insinuation was likely to be accepted as half-expected confirmation of deeply-rooted suspicions.

The best way to introduce the story of expanding access to the official wartime records is by considering the thrust and counter-thrust of attempts to document or demolish the fables that have shrouded Jubilee despite the best efforts of some very good historians. For example, when David Irving set out in 1963 to prove that the Germans had enjoyed foreknowledge of the intention to raid Dieppe, he was guilty, in the words of Captain Stephen Roskill, of 'the restatement of an already old and discredited illusion'. Their confrontation is still of some interest for its own sake, but it is especially instructive today as a period piece, a reminder of how limited the opportunities were in 1963 to work with official sources and how carefully the authorities guarded their files. Irving has since gone on to make a name for himself as the foremost exponent of viewing the war from Hitler's vantage point – at the cost, according to his critics, of occasionally pushing this approach to perverse lengths. Yet in 1963 he could still be described as 'a young historian', one with an undeniable knack for disinterring German wartime documents, whereas Roskill was the Royal Navy's Official Historian and thus, in the jargon of the day, an Establishment figure.[21]

Irving's first article appeared in the London *Evening Standard* on 1 October 1963, entitled 'Dieppe: Hitler *knew* it was coming'. The 'deadliest sand game in history' had been organized on 15 August 1942 at HQ Luftflotte 3 (3 Air Fleet), with an 'Admiral Mountbatten' in one room directing an assault on Dieppe, the chief Intelligence officer on the staff of C-in-C West directing its defence in another, and an umpire in a third. This 'macabre' game lasted until 17 August and was finally played out with the umpire, General Frölich, formerly Air Commander Africa (Fliegerführer Afrika) ruling that the Allies had been driven out of France in five days. That afternoon, General Ulrich Kessler, Air Commander Atlantic (Fliegerführer Atlantik), summoned a conference to his HQ in Angers at which he announced that the British were planning to attack Dieppe in the very near future. One German Air Force (GAF) unit, which had earlier been ordered to move from Rennes to Norway, left behind a full ground staff and a few aircraft to facilitate its speedy return in an emergency. 'Who told the Germans it was going to be Dieppe? Who has it on his conscience?'

The following day, the *Evening Standard*'s Bonn correspondent reported on the strength of an interview with Kessler that the GAF were in possession of 'the British naval code'. 'It clearly helped us to anticipate the raid well beforehand and take the necessary measures to meet it', Kessler was reported to have said. He added that he had always been in favour of attacking shipping whose positions were so easily ascertained rather than difficult inland targets in the UK.

Having aroused 'a tremendous controversy', according to the *Standard*, and having been dismissed as unreliable by an Admiralty spokesman, Irving published a second article on 14 October entitled 'Here is the Proof'. Much of his supporting documentation consisted of a '2,000 word telegram' that Hitler had despatched to his C-in-C some five weeks before Jubilee and which had been featured in some detail in his first article. The 'swift and crushing success' against convoy PQ 17 early in July left England with two alternatives: launch a Second Front soon or risk losing the Soviet Union as an ally. According to Irving's reading of the text, Hitler pointed out that evidence had been accumulating of preparations for a landing in the West. There had been a flood of reports from agents in Britain and from other Intelligence sources; a strong force of landing craft had been detected by aerial reconnaissance along the English south coast; and the RAF had been restricting operations in the last few days. The most threatened stretch of coastline was the Channel coast between Dieppe and Le Havre because it lay within range of fighter cover from England. This telegram was backed up by a 'sheaf of telegrams' ordering German reinforcements west. The 'mobile SS command' was transferred to France to take command of all SS troops there; Goering was ordered to move 7 Air Division and the Hermann Goering Brigade to France as well; and so on. How was it possible that this telling evidence had been missed by Roskill in the relevant volume of his Official History, *The War at Sea* (1956)?

Irving's answer was that his key documents had been released in the United States only in 1961, where he had found the telegram of 9 July in one of Admiral Dönitz's files. He backed this up with selections from Hitler's table talk and Albert Speer's record of a 'full scale conference' summoned by Hitler on 14 August to review French coastal defences. Even then, Irving concedes, Hitler was not quite sure *where* the Allies were going to attack. But this last piece of the jigsaw must have fallen into place in time for Kessler's conference on 17 August, thanks no doubt to the breaking of the British naval code. Irving then made the startling claim that, as he had first learned of Kessler's conference from an Allied Intelligence document of 1943, Allied authorities had therefore known since then that the Germans had been forewarned; but this document was still classified under the Public Records Act of 1958 and Irving had agreed to a suggestion by a 'senior Admiralty officer' not to 'disclose its identity'.[22]

Roskill pointed out at once in his rebuttal that Hitler's telegram of 9 July had long been familiar to him and that its text offered no support for Irving's conclusions. Hitler, for instance, had marked the southern part of the Dutch coast and the north coast of Brittany as threatened sectors, not just the stretch between Dieppe and Le Havre. He took issue with the reference to 'agents in Britain', because the original text merely referred generally to agents; 'other sources' indicated that the reports in question almost certainly did not come from the UK. The motorized division Grossdeutschland, which Irving included among the reinforcements moved to France in July, was not mentioned in the Order at all, far less did it leave Russia for the West. The staff at 3 Air Fleet had not been acting on the basis of specific intelligence: 'such anticipation regarding the enemy's possible intentions is, of course, a commonplace of tactical exercises'. The disclosure that the Royal Navy's codes and ciphers had been partially compromised hardly amounted to a 'revelation', considering that *The War at Sea* mentioned the fact at least three times. In the interests of scotching persistent rumours, Roskill thought it 'justifiable, even necessary' to add an Appendix to his article in the *Journal of the Royal United Services Institution* listing the principal documents on which his official version of Dieppe had been based.[23] *The War at Sea* had been published without footnotes to unpublished official sources, on the assumption that such sources would never be declassified anyway in the lifespan of its readers.

Roskill had the better of the exchange, although he was expecting too much in hoping that the issue would then be settled 'beyond all doubt'. Neither Irving nor Kessler produced the text of the intercepted naval signal(s) between the conference on 13 (not 14) August and 17 August identifying Dieppe as the target of a forthcoming raid. In any case, such an intercept was required even more promptly than that to allow the 'deadliest sand game in history' to start on 15 August. Roskill might have gone on to point out as well that Hitler's 'swift and crushing success' was actually plural in the original German and did not refer to PQ 17, and that 7 Air Division and the Hermann Goering Brigade had been in Normandy since the spring of 1942. More to the point, Irving had no business representing the 2,000-word telegram of 9 July as some sort of collector's item since Stacey had already taken account of it in his Official History of the Canadian Army in 1955. It belonged to a category of documents known as Führerbefehle (Orders issued by Hitler himself) and the complete text was published in 1963 as one of the documentary appendices ('Führerbefehl vom 9 Juli betr Verlegung von Waffen-SS-Verbänden in den Bereich des OB West', OKW/WFSt 551 213/42 gKdos Chefs) to Volume II of the German High Command (Operations) War Diary (*Kriegstagebuch des Oberkommando der Wehrmacht*

[*Wehrmachtführungsstab*]). Hitler's Orders were intended to be 'summary, imperative and immediate', according to H.R. Trevor-Roper in his edition of *Hitler's War Directives 1939–1945* (1964); Directives (*Weisungen*), on the other hand, laid down general guidelines and left it to the C-in-C in question to implement them.[24]

Irving's case would have been better served had Hitler issued his Order of 9 July a week earlier – which, by the way, he could well have done – and if Rutter had been launched on 5 or 6 July. Nevertheless, even this sequence would not necessarily have been conclusive proof of foreknowledge, as can be shown by considering a comparable case in March 1942. Hitler issued an Order on 23 March, in which he drew attention to the Brest and Cherbourg peninsulas as the likeliest places for the Allies to invade the continent; the islands off the south coast of Brittany between Brest and St Nazaire could, it was suggested, be of particular interest to the enemy. All available reserves were to move west of the line Caen–St Nazaire, and on 27 March 24 Panzer (Pz) division began to move west from Châlons-sur-Marne; map exercises were held in 7 Army area not unlike the one at 3 Air Fleet on 15 August. Then, on the night of 27/28 March the raid on St Nazaire took place. Yet, despite a far more direct and suggestive connection between one of the Führer's Orders and an impending British operation than Irving was able to establish, there has never been any question that the St Nazaire raid (Operation Chariot) achieved surprise, because Chariot had the smash-and-grab quality of a true raid and, unlike Jubilee, never lent itself to being encumbered with a heavy burden of speculation and suspicion. The move of 24 Pz, for example, took twelve days and 73 trains – 'too slow a rate to be regarded as a counter measure', according to British Intelligence.[25] Case closed.

So far as the release of classified wartime sources was concerned, the key measure was the British Public Records Act of 1967. The Act of 1958 had introduced the concept of statutory right of access to the public records but now they were to be released for public inspection after an interval of thirty rather than fifty years. The records began to be opened on 1 January 1970; however, to spare researchers the frustration of a piecemeal release of files occupying seven miles of shelving, the most significant military, political and administrative documents for the rest of the war were made available at one stroke in 1972.[26] Ronald Atkin, a British journalist, was the first non-official historian to make use, for a full-length study of the raid, of the wealth of material on Dieppe in the records of COHQ (DEFE 2), as well as in those of the three services (AIR, ADM and WO). Norman Franks filled in a long-standing gap in the history of the operation in *The Greatest Air Battle* (1979), using the same sources.[27] To understand the impact of the 1967 Act, all that is required is a comparison of Atkin's book with Terence Robertson's (1963). The latter has held up better than some of its reviewers evidently thought it deserved to; nonetheless, Robertson could not provide the solid underpinning and assurance that only access to official files can give.

Thanks to the enlightened policy introduced in 1967, there is no longer any excuse for mystery to surround the GAF map exercise on 15 August. Roskill hinted in his documentary appendix at the interrogation of a GAF officer captured in the Middle East, but without providing documentary references. What happened can now be followed in files DEFE 2/337 and DEFE 2/338. According to a report from AI(K) on 7 March 1943, 'an experienced GAF Oberleutnant' had been captured in North Africa and brought to the UK. He had been adjutant of III Gruppe, Kampfgeschwader 26 (III/KG 26), based the previous August at Rennes. Earlier in the summer, III/KG 26 had been on a training course at Grosseto in Italy; on or about 10 August, all but two of its Junkers (Ju) 88s were ordered to Banak in northern Norway to operate against the next PQ convoy, PQ 18, which had been reported off Iceland. The POW had stayed behind in charge of the ground defence of the base at Rennes and attended Kessler's conference at Angers on 17 August. Along with representatives of other GAF units in western France, he was treated to a recapitulation of the map exercise at 3 Air Fleet. A five-pronged attack was hypothecated, aimed at Dieppe itself; the weather on the chosen day was likely to be clear in the morning to enable the RAF to cover the landings, but cloudy in the afternoon when the GAF would be operating in full strength. So far, undeniably, there was more than a passing resemblance to Jubilee, given the landings on five beaches and the weather conditions that actually prevailed on 19 August. With talk of the consolidation of a bridgehead, however, verisimilitude begins to crumble. The Allies were expected to push along the coast towards Le Havre and inland as far as Rouen by the end of the second day. The following day, they were supposed to have reached Paris and to be poised to strike south into the Unoccupied Zone as if to split German forces in France in two. But somewhere near Beauvais they would be skewered on the tips of two German counter-attacks and driven back to the coast. Clearly this was more than a raid or even 'a reconnaissance in force'.

The interrogation report was nevertheless initially accepted by the senior staff at COHQ as confirmation of their worst fears about Jubilee. Major-General J.C. Haydon, Vice Chief of Combined Operations, minuted Mountbatten on 13 March 1943 that it was evident that the Germans had known about Rutter in advance, that they were aware of the decision to remount it, and that British Counter-Intelligence had been unaware of these breaches of security. Haydon's reference to Rutter was presumably based on the observation that III/KG 26 had been ordered to stand by for an emergency return to France while still in Italy in June. 'Had we known what now appears from the prisoner's statements', Haydon concluded, 'I am sure you would have ordered the cancellation of Rutter in the first place, and Jubilee in the second.' Naturally, the matter could not rest there and the full implications of the interrogation were investigated by ISSB. The Joint Intelligence Committee (JIC), to which the ISSB reported, submitted the results on 31 March. The POW was found to be correct in only two respects: in talking of a five-pronged attack, and one that was to take place 'shortly'. But the German choice of Dieppe as a suitable landing-place for their map exercise was not based on hard intelligence. 'There is in fact fairly conclusive evidence from the many sources of information received before, during and after the raid, that there was no leakage of information to the enemy.'[28] So Irving was right after all about the war game and the conference at Angers, but presumably only an official historian like Roskill could have had access to such sensitive material (in 1963–4) as the JIC report.

Irving's trip to the US to seek out German sources released in 1961 likely meant that he had managed to negotiate access to German naval records then held by the Office of Naval

Intelligence (ONI). The complete archive, covering both World Wars and the inter-war years, was captured at Tambach Castle near Coburg in 1945 and microfilmed by the Admiralty and ONI. The British gave the files PG reference numbers – standing, rather engagingly, for *P*inched from the *G*ermans – and copies of the thousands of reels of this invaluable source have been available at the National Archives in Washington since 1972, although only in 1984 did the first volume of a published guide appear to complement those already available for Military and Diplomatic records.[29] Since there is still only one guide for the Second World War (relating to U-boat warfare) and no logical correspondence at all between PG numbers and reel numbers (T 1022 series), working with these records is something like sampling a lucky dip. For example, reel T 1022/2364 contains PG 3447, which in turn contains the Abwehr report used by Roskill in 1964 to prove that Admiral Canaris's intelligence service had received no reports from its agents that the enemy were planning a raid on Dieppe. This document (Abwehrleitstelle Frankreich, B Nr 6860/42 Gkdos, Leiter III, 6.10.42) is indeed a lucky find, considering that the Abwehr destroyed records wholesale during the war and that intelligence-related material was usually combed from the files before their release. Some of the boxes in the T 1022 series are empty; others bearing the inscription 'sanitized' contain significantly short reels.

The Abwehr report offers a possible answer to what Stacey called one of the mysteries surrounding Jubilee: How did British Intelligence manage to locate the wrong German division in Dieppe? The Intelligence staff at GHQ Home Forces were confident that 110 ID had completed a move from Russia to France on or about 11 June; indeed, in the series of weekly Intelligence reviews, known as Martian Reports, which began to circulate that month, it was affirmed in No 7 of 14 July that 302 ID had left the Dieppe area and been replaced by an unidentified division. Infantry divisions in the 300–350 series had been raised in the winter of 1940–41 specifically for defensive service in the West and were known to the British as 'low category' divisions. Following the winter crisis on the Eastern Front, a number of low category divisions were known to have been withdrawn from France for service with the Field Army. In sectors where the Germans were not prepared to accept any reduction in strength on the coast, the low category division was replaced by a first-class, if battered, division from Russia, the first instance of this sort of exchange having led to the appearance of 106 ID at Calais. At any rate, it was thought extremely likely that 110 ID had replaced 302 ID, and the Jubilee Operation Orders referred to the former as the Dieppe garrison. The division had in fact never left Russia and 302 ID did not move east to meet its fate there until the end of December 1942.[30]

The Abwehr pointed out that there had been talk of 110 ID moving to France; a closer check of this item 'would be interesting'. It was possible, however, that the confusion might have been the result of a simple mistake in translation. According to the report, a 'wireless game' had been in progress, involving the playing back of messages to London on a transmitter not known to be under German control. In reply to London's question about what German division was in Dieppe, the agent meant to say that Army vehicles in the locality bore a divisional emblem consisting of a white church with a small tower. He used the French word 'Nef' for nave, which also means vessel or barge; the emblem of 110 ID was a white Viking ship.

Whatever the truth of this explanation – there is at least one other of perhaps lesser implausibility – Roskill's wish that the question of German foreknowledge should be settled 'beyond all reasonable doubt' went unfulfilled after 1964. Thoroughly erratic and disconcerting allegations about Dieppe continued to crop up. No theory was too far-fetched to be given a respectful hearing, such as the Sylvan Flakes soap advertisement featuring a 'beachcoat' from Dieppe which resurfaced in 1963.[31] One book (1966), which appears in most bibliographies dealing with wartime intelligence and sabotage, not only repeated the claim of German foreknowledge but managed to have the raid take place a year late; another (1991), has it taking place a day early. Mountbatten was unaccountably misquoted in the Canadian press on the twenty-fifth anniversary of Jubilee to the effect that the Germans had known about it four days in advance. And it is still possible to read brief accounts of the operation in books published in the 1980s in which almost every statement of fact is demonstrably wrong.[32] More interestingly, though, the entire history of the war against Germany has been recast since 1964, as can be shown simply by following up the implications of one or two of Roskill's remarks.

It is now clear, for example, why he went out of his way to correct Irving on the apparently trivial point of the origin of agents' reports about the forthcoming raid. Adding reports from agents in England was one of Irving's embellishments of the original text for which Roskill rightly took him to task. But only with the publication of Sir John Masterman's *The Double-Cross System in the War of 1939 to 1945* (1972) was it revealed that all Abwehr agents in the UK had been caught and either executed or turned into double agents. Roskill was presumably aware of this but bound by the Official Secrets Act not to acknowledge it publicly, beyond his vague reference to 'other sources'. Masterman had served as chairman of the Twenty Committee – hence XX – which cleared the information to be fed to the enemy by the likes of Tate, Mutt and Jeff, and Dragonfly, and his book originated as an MI 5 in-house manual on the running of double agents. About a hundred copies were run off on the Foreign Office press soon after the end of the war; and it was published by Yale University Press only after years of wrangling between Masterman and Whitehall.[33] Roskill was presumably not prepared tacitly to accept that a report from a XX agent had betrayed either Rutter or Jubilee.

There was of course no 'forthcoming' raid on 9 July, and the reports referred to by Hitler must have been inspired by rumours of a Second Front or by the increase in activity along the English south coast starting in May. Strictly speaking, those agents' reports had no bearing on Jubilee. The Abwehr document meets this objection, being dated 6 October, but it was the work of Abwehr III – 'Leiter III' – meaning the Counter-Intelligence branch: that is, it covered reports about assignments given to British agents on the continent by their controllers, only four of which remotely suggested a special interest in Dieppe. On 22 July, a controller in Switzerland wanted to know what troops were manning the coastal defences between Dieppe and Le Havre; but he was also interested in the coast between Cherbourg and Deauville and Biarritz and the Spanish frontier. Also at the end of July, an agent of Abwehrstelle Angers, possibly a German double agent, was asked generally about the bunkers being built along the French coast. The other two possibly relevant reports dated back to April. Abwehr HQ in Paris understandably found that the sum total of this information scarcely betrayed a British intention to raid Dieppe.[34]

One of the officers whose name appears on the distribution list of this document – 'Referat III F, Oberstleutnant Reile' – published his memoirs in 1962. Reile was never unduly modest about claiming credit for the achievements of Abwehr Counter-Intelligence in the West, particularly with respect to the invasion in 1944. If there was a plausible point to be made or score to be settled with regard to Jubilee, he would certainly not have missed it, but instead he related an odd story of how his counterpart in Holland, Giskes, deceived the British by making a pretended attempt to blow up the masts of the wireless station at Kootwijk early in August. Giskes, who ran the most successful of the Abwehr's double-cross operations, established what might seem a somewhat arbitrary connection between the Kootwijk assignment from London and the Dieppe raid. The destruction of the masts was supposed to disrupt communications with U-boats in the Atlantic but it was not explained how such a dislocation could have had a bearing on a cross-Channel raid, or why, if it was so important, it had not been laid on before Rutter. Reile's account was repeated in 1966 by Gert Buchheit, another old Abwehr hand, and may therefore be accepted as the best possible front Abwehr III could present in this context.[35]

German agents in England and elsewhere abroad were a different matter altogether. They were run by Abwehr I – 'geheim Meldedienst' – and therefore were not covered by this document. Predictably enough, the mischievous forces presiding over Jubilee's historiographical fate have acted to fill this void. By the time Irving published *Hitler's War* in 1977 he had discovered evidence that Roskill would have found much more difficult to disprove. The German Naval Staff (Operations) War Diary contains a reference to an agent's report in October 1942 which was credited with special significance as having come from the same agent whose report on 13 August had pointed to Dieppe as the target for an impending raid. In a letter to the *Daily Telegraph* of 2 September 1989, Irving again alluded to this report as proof of German foreknowledge. Moreover, although he did not point this out, the report in October had come from the south of England, which might be taken as presumptive evidence that the other one had as well.[36]

Secondly, where signal intelligence (Sigint) was concerned, anyone familiar with the historiography of the war since 1964 is bound to find Roskill's admission that he had made no fewer than three references to German success in breaking naval codes and ciphers a little disingenuous; at the very least, this is a massive understatement of the achievements of the German Navy's wireless B[eobachtungs]-Dienst.[37] Roskill was of course bound by the Official Secrets Act to say nothing about the immeasurably greater success enjoyed by the Government Code and Cipher School (GC&CS) in attacking German ciphers, just as Masterman was in 1972. Not until 1974 was the Ultra secret broken by Group-Captain F.W. Winterbotham, who revealed for the first time that Intelligence was the great British success story of the war. Ultra, strictly speaking, was merely a security classification above Top Secret, but the term has been loosely expanded to include the content of the intercepted signals, described as decrypts and usually enciphered on the Enigma machine. *The Ultra Secret* demonstrated the primacy of high-grade Sigint over all other sources of intelligence in terms of reliability, scope and volume. Only such a source, it became clear retrospectively, could have provided the insight into the inner workings of the Abwehr essential to the running of the XX agents. Once again, however, exponents of the devil theory of Dieppe have risen to the challenge of keeping up with changes in the historiographical climate. In the same letter to the *Daily Telegraph* in 1989, Irving referred to a '1945 monograph' later released by the US National Security Agency which pointed out that there was 'hard and fast evidence' that Ultra had revealed that the Germans had known 'as early as 12 August' about Jubilee.[38]

The Ultra Secret was of necessity written from memory, without access to the decrypts or other primary sources, with the result that Winterbotham made mistakes and indirectly encouraged speculation about such incidents as the air raid on Coventry on 14/15 November 1940. Presumably one reason why the Cabinet Office authorized the publication of an official history of British Intelligence during the Second World War was simply to set the record straight. At any rate, Volume 1 (1979) of Professor Sir

Harry Hinsley's magisterial work included an Appendix on the raid on Coventry, complete with footnote references, including some to sources that were still classified. No thought had ever been given, it should be remembered, to an official history of Intelligence in the First World War or in peacetime – in fact the authorities in Whitehall admitted to the peacetime existence of the Secret Intelligence Service (SIS or MI 6) only in 1980, after it had been in business for almost seventy years.[39]

Hinsley's second volume (1981) included an Appendix on 'Intelligence Before and During the Dieppe Raid', which concentrated on tactical intelligence. As a result of putting too much reliance on one source – aerial photographic reconnaissance (PR) – the planners and Intelligence staff at COHQ 'underrated the strength of the defences at Dieppe and the topographical difficulties that would be encountered by a landing there'. Stacey and Robertson had made similar observations, particularly with reference to the unsuitability of the main beach for landing tanks and the failure to detect the gun emplacements in the cliff-faces. Only agents on the ground could have reported on the caves, as well as the anti-tank guns wheeled out after dark to guard access to the streets leading from the promenade. According to Hinsley, however, MI 6 were handicapped at the time by an intensive security drive in the coastal zone. On the other hand, accurate data had been collected on the strength and rate of reinforcement of the GAF, but an experiment to exploit low-grade Sigint to provide current air intelligence during the operation was not a great success. The raid, finally, had achieved tactical surprise despite the chance encounter with the convoy that 'might have been avoided but for failures of communications'.[40]

The Appendix, in short, essentially added detail and authority to what was already generally known. Somewhat ironically, it fails to ring entirely true only in its treatment of the one slip in a published source that it set out to correct. Stacey had suggested that a report of the move by 10 Pz division to the Amiens area had an effect on the cancellation of Rutter; Hinsley took this report to be one from agent Bertrand in the Unoccupied Zone which was apparently received after Rutter had been cancelled and thus had no bearing on that decision or on the subsequent one to resurrect the operation. This is true for the cancellation, assuming that Stacey's anonymous report was one and the same as Bertrand's, but the impression given by the Appendix that 10 Pz's move west from Soissons was unknown until Bertrand's report on 13 July is contradicted by evidence that the division's move had been reported in time to affect Rutter in another important respect. The Martian Report of 23 June had already mentioned an unconfirmed report of this redeployment, which was interpreted as a logical step inasmuch as the Germans evidently expected trouble in either the Boulogne or Le Havre areas. Boulogne had recently been the target for a raid and Le Havre was an obvious choice as a base-of-operations port; both were within easy striking distance from Amiens. The COHQ Intelligence dossier for Operation Sledgehammer East contains an item to the effect that 10 Pz moved to the Amiens area in early June and that its HQ had since been reported at Flixecourt, halfway between Amiens and Abbeville, the source quoted being GHQ Home Forces on 4 July. On 5 July, Rutter was modified from a two-tide into a one-tide operation – meaning that re-embarkation would take place at 1100 rather than 1700. And on 6 July Lieutenant-General Montgomery, GOC South Eastern Command, emphasized that the raid should be kept as short as possible because 10 Pz was only 80 miles direct (he must surely have meant 80 km) from Dieppe; if postponed again, he added, it should be cancelled. The Appendix commented that the decision on 5 July was made because of the danger of giving the enemy 'the chance to reorganize on a scale that might prevent the withdrawal of the expedition's tanks'.[41] It is difficult not to accept that the threat posed by 10 Pz division was perceived to be the most significant part of that 'reorganization', whatever Bertrand may or may not have been reporting.

Hinsley refrains from commenting on the Notes on the planning of Rutter compiled by Roberts and Baillie-Grohman on 9 July. Much of their concern was directed at the failure to produce a joint appreciation on which the Outline Plan should have been based. As a result there was nothing to guide the Force Commanders engaged on detailed planning on such points as why the necessity for cover from E-boat attack had been disregarded. There were too many channels, they observed, for the receipt of intelligence, so that the Rutter planners did not always share intelligence and sometimes obtained it by chance; COHQ Intelligence was merely an extra link in the chain which slowed up transmission from source to user. Baillie-Grohman, for one, was not impressed by Wing-Commander the Marquis of Casa Maury, the Senior Intelligence Officer at COHQ, whom he referred to as a 'foreign grandee'. But there was a point to the Notes not perhaps fully appreciated at the time, because it appears that 5 July was the earliest that intelligence about 10 Pz's move to the Amiens area reached the Rutter Force Commanders, despite the fact that Casa Maury undertook to 'devote special attention' to the division's reported presence there at a meeting of the COHQ Examination Committee on 30 June. This point came up during discussion of a proposed diversion for Operation National but was not mentioned during earlier discussion of Rutter. A Rutter Naval Operation Order of 29 June still referred to the Germans as having only weak reserves in the interior of France – one infantry and two 'raw' armoured divisions near Rennes and Châlons-sur-Marne, which can only have meant 6 and 10 Pz respectively. Only on 5 July was information about 10 Pz's move passed on to the Rutter Force Commanders and C-in-C Portsmouth, with the rider that the shortest time for advanced elements of this division to reach Dieppe was only four hours. Thus, unforgivably, Rutter could have been launched on 3/4 or 4/5 July as a two-tide operation, without knowledge of this serious threat since 10 Pz's move to Amiens had been *confirmed* in the Martian Report of 30 June.[42]

The last outpost of the Secret War was the history of strategic deception. The final volume of the Official History of Intelligence – Sir Michael Howard's Volume 5 on *Strategic Deception* – was published in 1990, a few weeks after Hinsley and C.A.G. Simkins's volume on *Security and Counter-Intelligence* (1990), but only after an officially imposed delay of ten years. Doubtless it was thought that the publication of these two volumes might jeopardize methods still in use or run the risk of disclosing the identities of former agents. Possibly, as John Keegan has suggested, the government was also deterred by the remarkable lack of 'authorization from any political or military authority' which marked the creation of the first agency to co-ordinate deception (the W Committee) in January 1941. In any case, Howard mentions Jubilee merely in connection with subsequent German assessments of the possibility of further cross-Channel operations on the same or a larger scale later in 1942. Hinsley and Simkins, on the other hand, are primarily interested in the operation's bureaucratic context and reveal for the first time that a War Office evaluation of Jubilee was among the classified material that found its way into the hands of the British Communist Party. Perhaps it is safe to assume at last that there are no revelations from the shadowy world of agents and double-cross still to come. Sir Stewart Menzies, wartime 'C' or head of MI 6, told his brother before he died in 1968: 'I hope and pray nothing will ever come out.'[43] If this much has now 'come out', what possible justification is there left for another book about Dieppe?

For a start, there is Masterman's arresting comment: 'It is sad, but interesting, to speculate whether the Dieppe raid might not have been more successful, or at least less costly, if it had been effectively covered.'[44] This aside invites elaboration, to say the least, but Howard's volume was not pitched at that level of detail. There is still room for a sharper definition of the reasons why Rutter and Jubilee were denied a cover plan; also, of how the concept of a medium-sized raid fitted into the 'war of nerves' being waged in 1942 as an undertone to the thunderous agitation for a Second Front.

Another inviting area of investigation, which was barely mentioned in the Hinsley Appendix, except in the course of a long quotation from the Combined Report's summary of the Intelligence dossier of the town and its defences presented to the Rutter planners on 8 June, is radar. The Rutter and Jubilee Operation Orders provided for an 'RDF expert' to dismantle and remove as many parts as possible of the GAF Freya apparatus located on the west headland, but said nothing about radar in its primary role of detection – as a threat, in other words, to tactical surprise. Nor was radar an issue of any significance in any of the post-mortems on Jubilee, no doubt because the German reaction to the initial assault was thought to be consistent with the achievement of tactical surprise. The authors of the Combined Report echoed Hughes-Hallett's conviction that surprise had been enjoyed even after the collision with the convoy, and Stacey concluded in 1955 that the lowering point for landing-craft from the LSIs had been beyond the range of shore-based radar, quoting Field Marshal von Rundstedt's Battle Report ('Gefechtsbericht über Feindlandung bei und beiderseits Dieppe am 19.8.42', Ia 2550/42 gKdos, 3.9.42) in support.[45]

Actually, C-in-C West changed his tune 10 days later, after it had been brought to his attention that a GAF installation in Dieppe had followed the landing force intermittently, only for the Navy to insist on a false evaluation of the plot. The first published acknowledgement of the fact that this particular dog had, after all, barked in the night was in an anonymous article in a German magazine in 1958, which was in fact the work of Willi Weber, the officer in command of the radar unit in Dieppe at the time of the raid. Not until the publication of Mordal's book in 1962 did this feature of the operation reach a wider readership; in his review, Hughes-Hallett referred to Mordal's 'new, and I should think dubious, material about radar information said to be available to the Germans'.[46] In 1967, the former 'RDF expert' surfaced in the person of Jack Nissenthall (later Nissen), who evidently thought his achievements at Dieppe deserved more recognition than they had received. 'Everybody was decorated', the former Flight Sergeant later lamented, 'except me'. James Leasor's *Green Beach* (1975) was a dramatic account of this dangerous mission, and Nissen himself weighed in with *Winning the Radar War* in 1987. Although the Freya was not captured, Nissen credited his observation of its aerial array and his cutting of the telephone lines leading from the station with the success of radar counter-measures used on D-Day. In 1991 came two incongruous developments. First, the substance of Nissen's claims was openly challenged for the first time. Secondly, the Prime Minister was approached in an endeavour to have the 1949 order putting an end to the consideration of citations for bravery during the war waived so that Nissen could be awarded the Victoria Cross; he was not recognized at the time, according to a newspaper report, because of the highly secret nature of his work, although Mountbatten would have 'decorated him on the spot' had he known about the provision for Nissen's Canadian escort to shoot him should he be in danger of falling into German hands.[47] Aside from the permutations of this extraordinary sub-plot, however, many questions originally raised by Weber's article remain unanswered, questions about the German system of coastal surveillance and defence, about British intelligence about German radar, indeed, about British radar. Such questions could only be addressed by research in some depth into this technical aspect of the raid's context.

Hinsley's text makes it clear that another factor of major importance in Channel operations was Ultra. The Appendix, however, never mentions it at all, although including references to intelligence almost certainly derived from the GAF Enigma; and the two best-known popular histories of Ultra – Winterbotham's and Ronald Lewin's (1978) – passed over the Dieppe raid without comment. So much for the expectations of those who took it for granted that the history of the war would have to be comprehensively rewritten after the breaking of the secret in 1974. Fortunately, the release of wartime records did not come to an end in 1972. Even before the publication of Volume 1 of the Official History, the first batches of thousands of decrypts – or at least their English translations – were made available to researchers when the Public Record Office's (PRO) new building at Kew was opened in October 1977. Included were signals in the German Navy's Heimisch key (ZTPG), which was used by ships and shore establishments in home waters, including the Bay of Biscay and Channel, and which GC&CS had read 'with virtual currency' since its introduction in August 1941. The decrypts for 19 August 1942 recorded the Navy's reaction to the collision between Convoy 2437 from Boulogne and 5 Group and to the subsequent landings. But despite the speed with which they reached the Admiralty's Operational Intelligence Centre (OIC) in London, via the teleprinter line from Bletchley, they had little immediate operational importance, which is doubtless why they were not mentioned by Hinsley.[48]

There is a good possibility, however, that the decrypts in the ZTPG series are not the whole story, once the focus of attention, again, is shifted away from a close-up of the events of 19 August. Setting aside for the time being the question of the validity of Irving's source, it is still legitimate to ask how dependably Ultra could have been relied upon to raise the alert that the security of an operation had been compromised. And if the signals in the Heimisch key were being read with 'virtual currency', what did they disclose about coastal convoy traffic, in particular the sailing of the convoy from Boulogne to Dieppe? What precautions were taken to protect the security of the Ultra source, and did these inhibit the use of intelligence derived from it? For if Irving's claim has any legitimacy whatsoever, that would seem to be the most plausible explanation for official inaction.

The initial release of decrypts for public inspection in 1977 (DEFE 3: War of 1939–1945 Intelligence from enemy radio communications) marked the start of a process still under way. Increments to DEFE 3 have often been listed in the Keeper of the Public Records' Annual Reports, together with other newly opened classes, such as the Minutes of the ISSB (WO 283) in 1982 and the 35-foot run of WO 208 (Directorate of Military Intelligence) in 1983. At the same time, however, it is important to retain a sense of documents that have not yet been, and others that may never be, released. A few are retained by departments under Section 3(4) of the Act of 1958; others, including some of the reputedly most voluminous series of decrypts, have failed to appear at the PRO at Kew, possibly because of cutbacks and lack of resources. So far as policy governing hitherto classified material goes, the Foreign Secretary stated in 1978 that there would be no difficulty in opening the records of the Service Intelligence Directorates but that records relating to 'the methods by which this material was obtained' would remain subject to the provisions of the Official Secrets Act.[49] Many of the sources Hinsley listed at the end of Volume 3 (Part 2) have PRO class numbers and are open, while others also with class numbers apparently remain subject to extended closure – CAB 121, for instance, containing material on Rutter in CAB 121/364. Even further beyond reach are records without PRO numbers at all or even footnote references in the text of the Official History; these, according to Hinsley, are 'unlikely ever to be opened at the Public Record Office'. Among the absolute exclusions, it need hardly be added, are the domestic files of MI 5 and MI 6. As Professor D.C. Watt put it in his review of the final two volumes: 'So far as the present government is concerned, we, the professional historians and the general public alike, have had our lot.'[50]

In specific terms, then, reports from MI 6 agents like Bertrand, the traffic of the double agents run by MI 5, the decrypts of Abwehr Enigma messages dealing with this traffic, and an indeterminable amount of other material whose release is deemed a threat to national security will all remain under wraps, for the policy is unlikely to change with future governments. Finding a

German agent's report whose origin, transmission, interception and circulation (on both sides) can all be filled in with confidence will accordingly remain something of an event; and identifying reports from the double agents in England as they appear anonymously in German sources will also remain exceedingly problematical, unless, of course, they are among the very small number mentioned in the Official History. Consequently, there will continue to be an imbalance between the considerable inferences that might be read into a report like the one mentioning Dieppe at the end of October 1942 and the reliable evidence available to test those implications.

It should perhaps be pointed out, as well, that far more records have been lost or destroyed than remain classified. For example, many basic German sources for the war in the West, such as the Intelligence (Ic) files for 15 Army and Army Group D, are missing; other Ic series which have been preserved for 1943 and 1944 often exist for 1942 only as items on depressingly long disposal lists of documents destroyed by order. The first sentence in the 'Vorwort' to the OKW War Diary for 1942 announces that the original must be considered to have been lost. Nineteen forty-two, in fact, was scarcely a vintage year for the survival of records on either the German or Allied sides. There were also fewer to begin with than for later on in the war. For instance, the entry into the war of the United States and the establishment of large combined staffs greatly increased the circulation of paper, so that no student of Allied cover and deception can afford to ignore the generous amount of material opened up at the US National Archives – but of course this covers only the period from Operation Torch on.

Why, then, 'a documentary investigation'? The truth of the matter is that this is by far the most reliable approach for the historian of the Second World War working beyond a certain level of detail. A technical report on radar, say, compiled in 1942 is almost certainly bound to be more trustworthy than recollections fifty years after the event. Besides, the records are so voluminous that they have ways of unexpectedly yielding secrets seemingly beyond reach. Paradoxically enough, the historian often ends up working with fragmentary clues extracted from the bulk of files, which at least affords an opportunity to act out one of the profession's favourite self-conceits, namely, that of the historian-as-detective. If there is a risk of producing history of an excessively academic and impersonal stamp – John Keegan's 'inanimate landscape of documents' – that risk must simply be taken. After all, quite enough has been published about Dieppe over the years that was not nearly academic and impersonal enough. Apart from anything else, the documentary record in all its fallibility should always be the chosen ground on which to confront the 'myth-makers, muck-rakers, mole-hunters, and other nonsense purveyors'.[51]

There is admittedly a danger of falling into the trap of assuming that the only reliable information is contained in war diaries, committee Minutes, planning papers and the like, or that the official record should always be treated as gospel truth. For example, the Combined Plan for Jubilee (JJ One) stated that CCO would exercise his duties as laid down by the COS in COS(42)218, para 11, of 27 July 1942. An unpublished history of Combined Operations (CAB 106/3) compiled by the Amphibious Warfare HQ in 1956 (while Mountbatten was First Sea Lord) repeated with slightly more emphasis than necessary that the raid on Dieppe was carried out under this 'new directive'. J.R.M. Butler in Part II of Volume III of the Grand Strategy series in the Official History (1964) accepted that CCO was 'generally responsible for launching the operation' under the same rubric. But the terms of the directive did not authorize CCO to launch the operation, but rather called for the submission of an Outline Plan for approval by the COS before the Force Commanders began the detailed planning. The three Commanders and their staffs must already have been well into detailed planning on 27 July to have produced JJ One on 31 July. All might still have been well if the reference to approval of 'the future raiding operation' in the COS Minutes for 12 August had meant Jubilee, as everyone up to and including Hinsley took it to mean. According to Brian Villa, however, the reference of 12 August could not have been to Jubilee because, for one thing, the Vice-Chief of the Imperial General Staff (VCIGS), who was filling in for the CIGS at the meeting, was later indignant because he did not find out about Jubilee until after it had taken place.[52] Therefore Jubilee was an unauthorized action on Mountbatten's part, and Villa's book daringly rests on this one point like an inverted pyramid.

This is documentary research with a vengeance, a *tour de*

force. Villa is acting not so much as a detective as a prosecuting attorney. Although Ziegler refuses to concede on the point of authorization for Jubilee, Mountbatten's biographer admitted that his subject was not above rewriting history 'with cavalier indifference to the facts'. Villa would probably add that he was not above rewriting the raw material of history more or less as it happened. This is an issue that belongs to Jubilee's top-level political context. At the more mundane operational level, it is difficult not to be curious as to what other operation was approved by the COS on 12 August and why *it*, contrary to the normal practice, was not named in the Minutes. An answer can safely be left for the time being to 'come out' in what will doubtless be yet another book about Dieppe long after this one.

NOTES

1. Oberkommando des Heeres, Op Abt (IIa), 'Führerrede zum Ausbau des Atlantik-Walles am 29.9,' 3.10.42, T 78 317. RG 24, C 3, 13,746, War Diary, G Branch, HQ 2 Cdn Inf Div, Oct. 1941–Aug. 1942: Int 3-3-3, Leigh-Mallory to Roberts, 27 Aug. 1942. Hobart to Liddell Hart, 25 Aug. 1942, Box 1, 1/376, Liddell Hart Papers, King's College, London. Colonel C.P. Stacey, *Six Years of War, The Army in Canada, Britain and the Pacific* (Official History of the Canadian Army in the Second World War, Vol. 1, Ottawa, 1955), p. 389, provides a detailed table of Canadian casualties.
2. Stacey, *Six Years of War*; Jacques Mordal, *Dieppe: The Dawn of Decision* (trans. Mervyn Savill, London, 1962); Terence Robertson, *Dieppe, The Shame and the Glory* (London, 1962); Ronald Atkin, *Dieppe 1942, the Jubilee Disaster* (London, 1980); Norman Franks, *The Greatest Air Battle: Dieppe, 19 August 1942* (London, 1979).
3. Review by Thomas J. Allen, *Canadian Forum*, 42nd Year of Issue, Dec. 1962, p. 202: 'The story needs a Canadian writer. Robertson has been in Canada only a few years'. Nigel Hamilton, *Monty: The Making of a General, 1887–1942* (Toronto, 1982), pp. 546–58; Philip Ziegler, *Mountbatten, The Official Biography* (London, 1985), pp. 185–97. Brian Loring Villa, *Unauthorized Action: Mountbatten and the Dieppe Raid* (Toronto, 1989). For a sample of the controversy, see Ziegler's review in the *Spectator*, Vol. 264, 10 March 1990, pp. 31–33, and subsequent correspondence.

4. MG 30 E 463, Hughes-Hallett, Unpublished Memoirs, p. 188, National Archives of Canada.
5. WO 283/7, ISSB Agenda and Minutes, 15 June 1942; CAB 79/56, COS(42)54th Meeting(O), 21 June 1942.
6. *Daily Mail*, 4 July 1942 – copy in RG 24, G 3, 10,872, Files, HQ 2 Cdn Inf Div: 232C2(D 37). DEFE 2/549, 'Draft Notes on the Preparation and Mounting of Operation Rutter', Military Adviser to CCO [Col A Head], 8 July 1942. WO 283/7, ISSB Agenda and Minutes, 1 Aug. 1942.
7. Baillie-Grohman, marginalia on Lessons Learnt from Jubilee, n.d., GRO/28, Baillie-Grohman Papers, National Maritime Museum, Greenwich. DEFE 2/334, NOIC Newhaven to C-in-C Portsmouth, A/2/105, 29 Aug.1942.
8. DEFE 2/63, Operation Abercrombie, a Narrative: it should perhaps be mentioned that there had been an abortive attempt to launch Abercrombie on 19/20 April. Lord Lovat, *March Past* (London, 1978), p. 238.
9. Ernest Betts, *The Film Business, a History of the British Cinema 1896–1972* (London, 1973), pp. 185–6. DEFE 2/334, HQ First Canadian Army, Intelligence Report, 8-5-1 Ops/56-1-1 Int, 22 Sept. 1942, Appendix D.
10. Notes of talks with Beaverbrook [with Bruce Lockhart present], 26/27 Sept. 1942, 11/1942/82, File 11/42, Liddell Hart Papers; Bruce Lockhart was Director-General of the Political Warfare Executive (PWE) at the time. In addition to the ISSB investigation (see n. 26), Brig. H. Allen, DDMI(S), made a 'careful inquiry into the security aspects of the raid' – DEFE 2/579, Report No. 109 by Historical Officer [C.P. Stacey], 17 Dec. 1943.
11. This was certainly true of the only really sizeable cross-Channel raids before Jubilee: Abercrombie (21/22 April) and Bristle (3/4 June).
12. AIR 16/765, Questionnaires, P/126A/1/INT, 26 Aug. 1942; DEFE 2/546, Operation Rutter, Minutes of Meeting at COHQ, 5 June 1942. Baillie-Grohman, comment on Minutes of Meeting of 4 June, GRO/22, Baillie-Grohman Papers.
13. Alastair Buchan to Mountbatten, 23 Aug. 1972, B 61, Mountbatten Papers, University of Southampton.
14. DEFE 2/333, Operation Jubilee, Report of Air Force Commander, Covering Letter, 11G/S 500/98/Ops, 5 Sept. 1942; Naval Force Commander's Covering Letter, NFJ 0221/92, 30 Aug. 1942. David

Fletcher, *The Great Tank Scandal: British Armour in the Second World War, Part I* (London, 1989), pp. 136–7. WO 219/1934, Martian Report No. 19, 7 Oct. 1942. KTB A, Armeeoberkommando 15, Ia Eingehende Befehle, OKW, OKH, Heeresgruppe: 'Grundlegende Bemerkungen des Oberbefehlshaber West Nr 11' (auf Grund der praktischen Uberprüfung und Erprobung der in Dieppe erbeuteten engl Panzerkampfwagen), Ia/Höh Offz f Pz Truppen, Br B Nr 337/42, gKdos, 12.9.42, T 312 502.

15. RG 24, C 3, 13,747, War Diary, G Branch, HQ 2 Cdn Inf Div, Aug. 1942: Naval Force Commander to CCO, NFJ 0221/92, 16 Aug. 1942.
16. WO 106/4194, Operation Rutter, Naval Operation Order, 0221/65, 29 June 1942. AIR 16/764, Operation Jubilee, Outline of Operation, 10 Aug. 1942, JNO 3, Instructions to Destroyers; Operation Jubilee, Detailed Military Plan, JG One, 9 Aug. 1942, Appendix D, 'Assault and Occupation'.
17. WO 106/4197, Operation Jubilee, Naval Force Commander's Report, NFJ 0221/92, 30 Aug. 1942, Enclosure No. 7, 'Bombardment Report'. RG 24, C 3, 13,747, War Diary, G Branch, HQ 2 Cdn Inf Div, Aug. 1942: Operation Order No. 1, 'Cauldron', [No. 4 Commando], 14 Aug. 1942. DEFE 2/328, Report of Landing at Puits by Captain G.A. Browne, RCA [FOO], 18 Jan.1943.
18. RG 24, C 3, 13,747, War Diary, G Branch, HQ 2 Cdn Inf Div, Aug. 1942: Operation Jubilee, Air Force Appendix to Combined Plan, 11G/S 500/98/Ops, 13 Aug. 1942, Attachment 7, 'Instructions to Smoke Aircraft'. Air 16/764, Operation Jubilee, Report of Air Force Commander, 11G/S 500/98/Ops, 5 Sept. 1942, Covering Letter and Appendix D, 'Detailed Chronological Narrative'.
19. WO 106 4197 contains a copy of the Military Force Commander's Report of 27 August and a covering letter from DMO at the War Office to Lt. Gen. Morgan (WO 0.596, 28 Oct. 1942): '. . . I understand that you are unlikely to get much that is useful out of the Military Report. I am told that the Naval Report is an excellent one and it might well repay detailed study by your Sailor.' Morgan at that time was in command of 125 Force preparing for a possible post-Torch landing in Spanish Morocco. Col J.A. MacFarlane RAMC, 'Dieppe in Retrospect', *Lancet*, No. 6242, 17 April 1943, 498–500.
20. A. Robert Prouse, *Ticket to Hell via Dieppe: From a Prisoner's Wartime Log 1942–1945* (Toronto, 1982), p. 25.

21. Irving was apparently offered the opportunity by Lord Beaverbrook, proprietor of the *Evening Standard*, who never bothered to conceal his dislike for Mountbatten or his opinion that the raid had been a deliberate fiasco to convince the Americans of the folly of a Second Front in 1942 (A.J.P. Taylor, *Beaverbrook* (London, 1972), p. 538).
22. *Evening Standard*, 1, 2 Oct. 1963. On 6 Nov. 1963, the Hamburg magazine *Der Spiegel* published its version – 'Hitler wüsst alles' which contained a number of elementary errors: Kessler, for example, was identified as Operations Officer (Ia) on the staff of C-in-C West.
23. Captain S.W. Roskill, 'The Dieppe Raid and the Question of German Foreknowledge', *Royal United Services Institution Journal*, CIX, No. 633, 1964, 27–31.
24. Stacey, *Six Years of War*, pp. 350–51; *Kriegstagebuch des Oberkommando der Wehrmacht (Wehrmachtführungsstab)* (5 Vols, Frankfurt-am-Main, 1963), II (2), pp. 1280–1; H.R. Trevor-Roper, *Hitler's War Directives 1939–1945* (London, 1964), pp. 23–4.
25. Hitler anticipated many of the provisions of the Order of 9 July in the last week of June following the aerial reconnaissance of 20/23 June (Ch. 3). WO 219/1933, Martian Report No. 4, 23 June 1942.
26. *11th Annual Report of the Keeper of the Public Records on the Work of the Public Record Office and Eleventh Report of the Advisory Council on Public Records*, Appendix One (House of Commons Paper 278, Session 1969/70).
27. Norman Franks, *The Greatest Air Battle: Dieppe, 19 August 1942* (London, 1979).
28. DEFE 2/337, Haydon to Mountbatten, 13 March 1943. D.C. Duccan to Group Captain Willetts, COHQ, JIC/407/43, 31 March 1943, refers to the ISSB report on the matter. AI(K) was the section of Air Intelligence responsible for POW interrogation.
29. See Robert Wolfe (ed), *Captured German and Related Records: A National Archives Conference* (Athens, Ohio, 1974). See also F.H. Hinsley and H.M. Ehrman (eds), *A Catalogue of Selected Files of the German Naval Archives Microfilmed at the Admiralty, London, for the University of Cambridge and the University of Michigan, Project No. 2* (1964).
30. WO 219/1933, Martian Report No. 7, 14 July 1942. DEFE 2/330 contains a dossier from GHQ Home Forces (HF/INT/222/2/40, 6 Aug. 1942) to COHQ Intelligence on the history of 110 ID.

31. The Sylvan Flakes soap-powder advertisement mentioning a 'beachcoat from Dieppe' appeared in the press just a day or two before the raid and was investigated by MI 5 (WO 238/6, ISSB Agenda and Minutes, 2 Nov. 1942); it cropped up again in the *Daily Sketch*, 4 Oct. 1963 (George Begley, *Keep Mum! Advertising Goes to War* (London, 1975), p. 79); and was most recently mentioned in Paul Fussell, *Wartime: Understanding and Behavior in the Second World War* (New York, 1989), pp. 36–7.
32. E.H. Cookridge, *Inside SOE: The Story of Special Operations in Western Europe 1940–1945* (London, 1966), pp. 262–4, not only dates the raid in 1943 but associates it with events that took place that summer. Richard Lamb, *Churchill as a War Leader – Right or Wrong?* (London, 1991), p. 174. Toronto *Daily Star*, 21 Aug. 1967: 'Germans were ready for Dieppe Raid: Mountbatten'; Robert Goldston, *Sinister Touches: The Secret War against Hitler* (New York, 1982), pp. 146 ff.
33. For Masterman's own account of his struggle to publish *The Double-Cross System in the War of 1939 to 1945* (New Haven, 1972), see his autobiography, *On the Chariot Wheel* (Oxford, 1975), Ch. XXXII, pp. 348–61.
34. Abwehrleitstelle Frankreich, B Nr 6860/42 gKdos, Leiter III, 6.10.1942, T 1022 2364.
35. Oscar Reile, *Geheime Westfront: Die Abwehr 1935–1945* (Munich, 1962), p. 427; Gert Buchheit, *Der deutsche Geheimdienst, Geschichte der militärischen Abwehr* (Munich, 1966), pp. 340–41: 'Zweifellos hatte damals der britische Nachrichtendienst versucht, den Funkverkehr von Kotwijk im Hinblick auf den geplanten Überfall auf Dieppe lahmzulegen.'
36. Irving, *Hitler's War* (2 vols, New York, 1977), I, p. 446 (footnote: II, p. 948). Kriegstagebuch der Seekriegsleitung: 1 Abteilung, Teil A, 20.10.42, T 1022 1677.
37. Roskill, 'The Dieppe Raid and the Question of German Foreknowledge', *Royal United Services Institution Journal*, CIX, No. 633, 1964, 27. See, for comparison, Heinz Bonatz, *Seekrieg im Äther: Die Leistung der Marine-Funkaufklärung 1939–1945* (Herford, 1981).
38. F.W. Winterbotham, *The Ultra Secret* (New York, 1974); *Daily Telegraph* 2 Sept. 1989.
39. F.H. Hinsley *et al.*, *British Intelligence in the Second World War* (5 Vols, London, 1979–90), Vol. 1, Appendix 9: 'Intelligence in

Advance of the GAF Raid on Coventry, 14 November 1940', pp. 528–48; *The Times*, 12 May 1980: '69-year fiction over MI 6 comes to an end'.
40. Hinsley, *British Intelligence*, Vol. 2, Appendix 13: 'Intelligence Before and During the Dieppe Raid', pp. 695–704.
41. Stacey, *Six Years of War*, p. 339; WO 219/1933, Martian Report No. 4, 23 June 1942; DEFE 2/561, Operation Sledgehammer East. RG 24, C 3, 13,611, War Diary, GS Branch, HQ 1 Canadian Army, April 1942–Feb. 1943: Memorandum of a discussion between McNaughton and Montgomery, 6 July 1942, concerning Rutter, 8-3-4 Ops, T 6672.
42. AIR 16/760, Operation Rutter, 'Notes of Staff and Planners', 9 July 1942; DEFE 2/2, War Diary COHQ, Minutes of Examinations Committee Meeting, 30 June 1942; WO 106/4194, Operation Rutter, Naval Operation Order, 0221/65, 29 June 1942; DEFE 2/542, Rutter Intelligence, I/67, 5 July 1942. Baillie-Grohman, comment on Minutes of Meeting of 4 June, GRO/22, Baillie-Grohman Papers: 'I recognised him [Casa Maury] as a polo player'. WO 219/1933, Martian Report No. 5, 30 June 1942. The Appendix merely alludes to the criticism (Robertson, *Dieppe*, p. 115) that COHQ Intelligence was slow and inefficient and avoids altogether the question of the suitability of Casa Maury as Senior Intelligence Officer (SOI). Casa Maury, a former manager of the Curzon Cinema, was eased out early in 1943 when the Air Ministry refused CCO's request to promote him to Group Captain (See correspondence with Layton, CAS etc. in B 26, Mountbatten Papers).
43. Michael Howard, *British Intelligence in the Second World War: Volume 5, Strategic Deception* (London, 1990), pp. 52–8; F.H. Hinsley and C.A.G. Simkins, *British Intelligence in the Second World War: Volume 4, Security and Counter-Intelligence* (London, 1990), p. 285. John Keegan, *Weekend Telegraph*, 1 Dec. 1990. *The Times*, 1 June 1977: 'Disclosures "would have horrified MI 6 chief"'.
44. Masterman, *Double-Cross System*, p. 108.
45. Confidential Book 04157 F; Stacey, *Six Years of War*, p. 360.
46. KTB A, AOK 15: 'Grundlegende Bemerkungen des Oberbefehlshabers West Nr 10', Ia Nr 2676/42 gKdos, 13.9.42, T 312 502. W . . ., 'Das Unternehmen.. Jubilee, Die Landung der 2 kanad Division am 19 August 1942 bei Dieppe, Gefechtsaufklärung durch Radar', *Der Deutsche Soldat*, Heft 4/1959, 117–19 (the identity

of W was confirmed by Willi Weber in a letter to the author, 31 Aug. 1982). Hughes-Hallett, *Spectator*, Vol. 211, 23 Aug. 1963.
47. Nissen quoted in *Toronto Star*, 21 Nov. 1991. James Leasor, *Green Beach* (New York, 1975). Jack Nissen, *Winning the Radar War: A Memoir by Jack Nissen and A.W. Cockerell* (Toronto, 1987). J.R. Robinson, 'Radar Intelligence and the Dieppe Raid', *Canadian Defence Quarterly*. Vol. 20, No. 5, 1991, pp. 37–43. *Daily Telegraph*, 7 Oct. 1991: 'VC urged for scientist in Dieppe raid'.
48. Ronald Lewin, *Ultra Goes to War: The Secret Story* (London, 1978). The decrypts are in DEFE 3/187. Hinsley, *British Intelligence*, Vol. 2, Appendix 4: 'Enigma Keys attacked by GC and CS up to mid-1943', pp. 658–68.
49. *25th Report* (1983), H C 583, Session 1983–4; *26th Report* (1984), H C 550, Session 1984–5. Hinsley, *British Intelligence*, Preface to Vol. 3 (2), p. x.
50. Hinsley, *British Intelligence*, Vol. 3 (2), pp. 961–5; *Sunday Times*, 5 Aug. 1990.
51. John Keegan, *The Face of Battle: A Study of Agincourt, Waterloo and the Somme* (London, 1976), p. 33. D.C. Watt, *Sunday Times*, 5 Aug. 1990.
52. CAB 106/3, History of Combined Operations Organisation 1940–1945, Amphibious Warfare Headquarters, 1956, p. 40. J.R.M. Butler, *Grand Strategy*, Vol. III, Part II, (London, 1964), p. 639. Hinsley, *British Intelligence*, Vol. 2, Appendix 13: 'Intelligence Before and During the Dieppe Raid', p. 696. Villa, *Unauthorized Action*, pp. 44–5.

2

'Spy Stuff' – Agents, Cover and Deception

Sometimes it is possible to explicate minor historical puzzles with the precision and economy of mathematics. A case in point is provided by such scattered references as have survived to reports turned in by one of the Abwehr's agents in Paris. According to a telegram of 22 January 1943 from the Abwehr station in Hamburg (Abwehrstelle or Ast X) to Berlin (Abwehr I H West), 'Zuv[erlässig] A 3924' had foretold on 19 January a large-scale American operation against the west coast of France in about 14 days. This was only one of a suspiciously large number of reports conveying roughly the same information. Possibly for that reason it was forwarded with some reservation, but not without the reminder that the agent (or V[ertrauens]-Mann) had given 14 days' forewarning of the raid on Dieppe.[1]

Ast X included the reference number of A 3924's Dieppe report (FS Ast X Nr 895/42 I H geh) of 4 August, which by some fluke of documentary survival is to be found in the records of its sub-station (Abwehrnebenstelle or Nest) in Bremen on reel T 77 1539 at the National Archives. On 29 July, apparently, A 3924 had reported that pro-British groups in Paris were showing unusual interest in the Le Tréport area, adding that the north of France would be the easiest part of the country to win over in support of an enemy landing.[2] Considering that there was no mention of Dieppe or 19 August, this item looks like a not-very-near miss chalked up as a bull's eye. Nor was it the only retrospective claim made with respect to Jubilee. Besides the report earlier referred to of 29 October, there were at least two other documented examples, one from an agent in contact with the Military Attaché in Berne and a second from a V-Mann in Lisbon, R 3764. R 3764 was ranked as 'reliable and proven', having uncovered the concentration of shipping at Gibraltar before the North African landings as well as the intention to raid Dieppe. Neither of these reports can be evaluated with quite the same assurance as A 3924's. Neither singly nor collectively and despite being described as 'timely', had they any documentable impact on the naval and military authorities in the West: only A 3924's was signalized by a brief mention in the Naval Staff (Operations) War Diary (KTB 1 Skl) on 6 August.[3]

Jubilee was far from the only operation of which the Abwehr claimed to have provided timely forewarning; indeed, for all large-scale Allied landings there seemed to be at least one agent's report later confirmed by events. In the case of Torch, for example, A 3924 predicted on the very day of the landing, 8 November 1942, a seaborne attack on Casablanca between 13 and 17 November; another report (R 3917) pointing to an American landing at Casablanca was received on 4 November. Hinsley observed that the enemy received a number of agents' reports before Torch that came 'uncomfortably close to the truth'. Fortunately for the Allies, however, intelligence from other sources such as Sigint gave them the confidence to discount intercepted Abwehr reports that might otherwise have suggested that the cover plans for Torch had been compromised. The German Intelligence machine was 'incapable in the absence of reliable intelligence [such as was provided the Allies by Ultra] of discriminating between the many rumours and reports it was receiving about enemy intentions'. As a result, General Alfred Jodl, Chief of the OKW Operations Staff, berated Admiral Wilhelm Canaris, Chief of the Abwehr, for having 'landed us in the soup again' over Torch; Canaris had already been dressed down by Hitler on 30 June after the fiasco of Operation Pastorius. By the early months of 1943, consequently, Canaris's somehow diffident, overstaffed and distinctly suspect organization needed all the credibility it could muster in view of its record of failure and the mounting challenge from competing Intelligence organizations such as Amt VI of RSHA.[4] The Abwehr's role was largely limited to the procurement of intelligence and its officers put their faith in volume of reports. Therefore, allusions to the accuracy of an agent's earlier reports probably signified more than simply the drawing of attention to the possible significance of his or her current offering: implicit at least was a grudge against the agencies responsible for evaluating the Abwehr's output.

How important were agents as a source of intelligence for the German High Command? Generally speaking, they were ranked well behind other sources in reliability and usefulness. Lieutenant-Colonel Ulrich Liss, chief of Foreign Armies West until 1943, and Gehlen, his counterpart in Foreign Armies East, both valued Sigint as their most important asset. The OKW War Diary retrospectively agreed on the primacy of Y (wireless) intelligence, and listed aerial reconnaissance and POW interrogation ahead of agents. German documents are replete with reminders of the need to treat unconfirmed reports from the Abwehr and diplomatic and Attaché sources with caution because of the obvious danger of disinformation. Liss, for instance, was careful to label reports of the presence of US troops in southern England as unconfirmed until Y intelligence established the presence of the first US division to arrive there in October 1942. War diarists often went out of their way to denounce an agent's report in ringing terms – *recht wenig vertrauenerweckend* – and von Rundstedt's weekly appreciations of the situation in the West, which began in August 1942, frequently expressed scepticism about often contradictory reports from England and elsewhere. Yet the Germans faced a special problem in gathering intelligence from across the Channel. There was little or no yield from POWs or captured documents, sources that stood the German Army in good stead in other theatres of operation; the Y service concentrated on order of battle intelligence; and already in 1942 aerial reconnaissance was spotty and becoming increasingly hazardous. Consequently, the Abwehr, which had started the war with only half a dozen agents in Britain, was under mounting pressure in the summer of 1942 to make the most of those it had managed to infiltrate there; and the desperate shortage of intelligence from across the Channel led to a predisposition to accept the reliability of what little could be collected.[5] Agents, after all, were still the best bet to disclose enemy intentions as to the timing and targets of operations.

Individual reports almost never achieved the documentary distinction of a mention in the OKW War Diary. One that did, on 19 May, briefly stated that Vice-Admiral Mountbatten was planning a landing operation. The Naval Staff War Diary for the same date had 'Lord' Mountbatten embarking at Dunfermline on 15 May with 1100 Commandos and eight to ten battalions of Scottish Command for an unknown destination; another entry on 23 May put Mountbatten and his force at Lerwick in the Shetlands – 'interesting, to say the least', according to the Naval Operations Staff.[6] There had already been a large number of rumours and reports about an impending invasion of Norway and Denmark, and a state of 'Alarm Readiness' had been declared for all three services in Norway on 19 May following a report on 17 May warning of landings all along the coast. Some of these reports were generated by Operation Hardboiled, a notional landing on the Norwegian coast to capture Stavanger and its adjacent aerodromes. Next to nothing of Hardboiled's documentary record has apparently survived but a leading feature of the operation was an exercise in the Scottish Highlands with mountain troops (Operation Bess), supported by press reports of the presence of Norwegian interpreters in Scotland, the translation of billeting requisitions into Norwegian, and so on. According to the Official History of Deception, the double agents passed on some of the rumours circulating in London about a landing in Norway. Among the numerous Abwehr reports, one received on 25 May to the effect that the troops taking part in the mountain training had returned to Aberdeen and the Norwegian interpreters been stood down was obviously consistent with the Hardboiled 'story'; it came, moreover, by wireless from England.[7] The 'Lord Mount Batten' report, on the other hand, was credited to source 'ABC . . . aus London', most likely part of the 'network' of an inventing or uncontrolled agent in the business of selling the Abwehr plausible information picked up from such diverse sources as the British press, Hansard, railway timetables, reference books and the like.[8] The Abwehr had the habit of prompting agents for additional information about a development they had got wind of, and inventing agents were only too ready to oblige, though in this case Dunfermline, not being on the coast, was a notably bad choice as a port of embarkation.

Thus the Germans seldom received XX reports in support of British cover and deception operations without an accompaniment of distortion and embellishment. Hardboiled coincided with a major invasion scare in May that extended well beyond Norway. Von Rundstedt put his forces in France and Belgium through a large-scale exercise/alert over the Whitsun weekend. French railway workers, for example, were on standby to handle the possible movement of strategic reserves.[9] More-or-less unsupported agents' reports, in other words, could evoke a reaction from the German defensive system, and there was nothing to prevent a *Next of Kin* scenario from being acted out. In France, the Abwehr several times during the summer produced warnings of an enemy attack that were promptly acted upon by the military authorities. Twice in June, 302 ID spent a night in a considerably higher state of readiness than on 18/19 August. On 12 June LXXXI Corps warned that an Abwehr contact among French civilians had passed on information that the British were preparing to land somewhere that night under cover of bad weather. Another warning came on 25/26 June, forcing the infantry manning the coastal strongpoints to spend another uncomfortable night at Action Stations (Gefechtsbereitschaft), while the divisional artillery stood by with guns and limbers hitched up. On 18 July, an agent's report from Rouen predicted a landing at dawn in the Bay of the Seine; although the Abwehr passed this one on with some misgiving, an alert was issued anyway, despite very unsuitable weather conditions.[10] These reports – of local origin and betraying little concern for the future credibility of the source – were of course not the only incoming reports in those weeks but they were the ones that elicited a reaction from the military.

After Jubilee, the Germans never seem to have considered the possibility that the British had shot their bolt for the time being and therefore that the risk of reducing strength in the West could safely be run. Von Rundstedt issued a warning on 21 August that the enemy could be expected to launch further raids to regain his prestige. MI 14 informed the CIGS on 31 August that a Most Secret Source indicated that the Germans expected Allied raids to test their defences in the near future, 'although one reliable report states that they do not anticipate an invasion of the continent this year'. According to the Naval Staff War Diary, a small number of reports were now being received from London or elsewhere in England, several of which spoke of troop concentrations in the Isle of Wight area and of intentions to launch another operation in the Channel along the lines of Jubilee. Buried in the increasing 'noise', of course, were XX reports conveying the substance of Overthrow and Solo One, two of the cover plans for Torch aimed across the Channel and at Norway respectively. An 'Abwehrbericht aus London über Spanien' of 29 September claimed that there were ten divisions in Scotland earmarked for Norway; another, from 'a proven source', mentioned the start of construction at the end of September of bombproof, mostly underground 'invasion storage depots' between Shoreham and Swanage – similar reports were to become a feature of XX reports in 1943 at least. The XX agent Dragonfly was asked on 18 October about any possible operations and replied: 'Several soldiers believe soon attack like Dieppe, but bigger, against the north coast of France'.[11] According to Jodl's deputy at OKW (Operations), Lieutenant General Walter Warlimont, Hitler himself never read agents' reports at all and referred to them only to support his flashes of intuition. Whatever the truth of this, Hitler cited the increasing number of agents' reports early in October as the reason for ordering a higher degree of readiness in the West, and on 9 October moved mobile reserves into a better position to cover Cherbourg. Von Rundstedt considered a landing on 12 October a good possibility but still held back from assigning it the highest priority.[12] However, Army units and formations in Normandy and Brittany spent much of October and November on the *qui vive*.

Quite why some reports were singled out for inclusion in war diaries and others not is difficult to determine. Reports later cited as having provided accurate forewarning of an Allied operation more often than not turn out not to have been mentioned in the same source on or near the date in question. Yet some manifestly absurd items found their way in. The Abwehr roughly sorted reports according to the agent's track record and then deposited them with the staff of the Foreign Armies branches of OKH. The work of collating and evaluating all incoming intelligence about the Western Allies was done by Foreign Armies West, in the light of the strategic situation and any details that might be considered particularly relevant. After the parachute raid on Bruneval, for example, close attention was always paid to any mention of airborne forces. Foreign Armies West then passed on its daily situation report and bulletins, including selected raw reports, to the operational staffs, filtering from army to corps to division.

To return to the report, or rather reports, cited by David Irving in the *Daily Telegraph* on 2 September 1989. The Naval Staff War Diary, according to Irving, 'refers in October 1942 and March 1943 to "the *Abwehr* agent who reported the Dieppe raid" '. In fact, the report on 15 March was the one from the agent in touch with the Military Attaché in Berne – a different agent altogether. The Diary also carried R 3764's report on 15 February. But the entry on 31 October was undeniably special because only occasionally did the diarist flag an item as having come from the British Isles:

> According to Abwehr report from south of England of 29/9, an operation similar to the action at Dieppe is to be expected in the coming month. Abwehr adds the comment that same, extremely reliable agent, to whom the report in question can be traced, also gave forewarning of the Dieppe operation on 13.8.[13]

Here is an item rich in speculative potential, complete with what may have been a typing error for '29/10', since there would have been no point in drawing attention to a warning for the 'coming month' of October on its last day. On the other hand, the report may indeed have given a month's advance warning, since it was grouped with a 'further Abwehr report' of 16 October from London via Spain to the effect that warships and transports should be concentrated in the Thames estuary by 10 November at the latest.

That Jubilee's security had been breached is only the most obvious of a number of possible deductions. Inspired by the recollections of former members of Ast X about 'passing references' and the like from their master-spy in England to a raid in the Dieppe–Fécamp sector in mid-August 1942, Günther Peis, a West German journalist, spent some fifteen years tracking down V-Mann 3725. After some convoluted research, he traced 3725 to Watford, only to discover that he was none other than Tate, one of the stars in the XX firmament. This meant that when Peis published his book in 1976 he had to contrive some other explanation for what he took to be evidence of forewarning of Jubilee. He found it in betrayal, in the supposition that Tate or the previously mentioned Dragonfly had been used to pass on the report of 13 August, and possibly other accurate and timely reports about Jubilee, so as to build up his credibility in the eyes of the Abwehr. Once elevated to a new level of apparent trustworthiness, they were used to play a decisive part in putting across the strategic cover and deception for Torch.[14]

This conspiratorial hypothesis is so outlandish as to be perhaps beyond refutation by conventional methods, even if access to all the relevant documentation were open. But the circumstantial evidence against it is overwhelming. The main theme of Masterman's book, after all, is that of the patient and painstaking cultivation of the XX system with a view to its eventual use for the cross-Channel invasion in 1943 or 1944 – the decisive Allied operation of the war. Stunts involving a serious risk of blowing agents before D-Day were not part of the XX repertoire. Secondly, as Ewen Montagu pointed out, there was no need to enhance the credibility of Tate and Dragonfly or any of the other XX agents since they were already known to be in good standing with the Abwehr. Thirdly, the fact that Dragonfly's notional assignment to the Isle of Wight area, supposedly to discourage the Abwehr from attempting to set up new sources near the Combined Operations training and operational base opened there in May, was not carried through rather casts doubt on his authorship of the report of 13 August. Lastly, according to Nigel West, the unofficial historian of MI 5 who has a wealth of contacts in the Intelligence community, both Tate and his case officer have denied sending even passing references about Jubilee to Hamburg. Tate would in any case have been a poor choice for this purpose since his traffic, unlike that of agents controlled from Madrid and Lisbon, passed by landline between Hamburg and Berlin and therefore could not be read by GC&CS.[15]

Two other spy yarns should be mentioned in passing, though neither relates directly to the report in question. Ladislas Farago held that Jubilee was the occasion of 'the worst security leak of the war', although not with reference to the report of 13 August. Instead, he pointed to 'Princess' (A 1392 to the Abwehr), whose husband at the time was a British Embassy official in Lisbon. Although his publishers apparently found it prudent to omit the passage on Dieppe from *The Game of the Foxes* (1971), Farago produced no convincing evidence of a security lapse in a series of articles he did publish in the *Daily Express* in October 1971. There is surely something wrong with a version of Jubilee that sees some connection between it and the 'Dunfermline' Abwehr report of 15 May and requires the British Military Attaché in Lisbon to have enjoyed foreknowledge of the operation which even the JIC in London had not. Over and above that, the Ultra secret was still a secret in 1972, so that neither Farago nor Hughes-Hallett, who remonstrated with the editor of the *Daily Express* about the 'lady spy', could have known how comprehensively communications between Berlin and the Iberian peninsula had been compromised since the breaking of the Abwehr Enigma key in December 1941. According to Hinsley, Abwehr messages were being read by GC&CS at the rate of 'some 3,000' *a week* in the first half of 1943.[16] To postulate that an agent capable of pulling off the greatest security leak of the war could have continued to operate within the diplomatic circle in Portugal for any length of time is beyond belief. Yet Farago remarked that Princess's treachery went undetected; she was still supposedly passing accurate intelligence to the Germans in the summer of 1944.

It was left to Leonard Mosley to rise to the challenge of coming to terms with both the XX system and Bletchley Park's reading of Abwehr signals by conjuring up an agent operating outside the control of the Abwehr or MI 5, whose signals either escaped the notice of the Radio Security Service or were unreadable by GC&CS. This agent – the Druid – was sent to England to check on the reliability of the Abwehr's networks on behalf of the rival RSHA. The Druid found out about both Rutter and Jubilee from a disgruntled French-Canadian officer while touring the south of England with a concert party. Unfortunately, *The Druid* (1981) is devoid of footnotes and bibliography, and Mosley offers no explanation of how he came by the still classified Minutes of the XX Committee, in particular those for a meeting presided over by Colonel T.A. Robertson, who denied that the meeting had ever taken place. Nigel West questioned the idea of a super-agent on the loose and pointed out a number of errors and inaccuracies, including the confusion of XX agent Garbo with the Spanish journalist Calvo. Operation Torch, it might be added, did not take place in October 1942 and the German High Command was not the OHL, or at least not during the Second World War. Mosley had two main oral sources: one later denied his contribution; the other was the traitor Philby. The latter was never a member of MI 5 or a liaison officer between MI 5 and MI 6, although he is represented in both those capacities in the book; nor was he above spreading mischief and disinformation through his contacts with journalists in the West. Sir Michael Howard in the Official History finds no reason to challenge the official claim made by Robertson in a memorandum of 15 July 1942 'that the only network of agents possessed by the Germans in this country is that which is now under the control of the Security Service' – a claim that ruled out undetected agents as well as the possibility that the double agents might be obvious decoys to conceal the activities of a network of genuine spies.[17] Understandably enough, veterans of the wartime Security Service petitioned the publishers, Eyre Methuen, to reclassify *The Druid* as fiction.

Besides a striking level of implausibility, these three accounts have something else in common. It has never been enough for proponents of the breach-of-security explanation of Jubilee merely to uncover a possible leak: that leak must contribute directly to the disaster on the beaches at Dieppe. Unfortunately for this hypothesis, there is no registration of any alleged 'special alert' for 302 ID in sources where it would certainly have been recorded. For example, the War Diary of 15 Army shows that its commander issued an address to his troops on 18 August to congratulate them on having just survived the period of imminent danger – *drohende Gefahr* – that had started on 10 August. The enemy must have realized the hopelessness of attacking the coast held by 15 Army, he trumpeted; everyone, including the Todt construction workers, could now enjoy a few days of exercise, sleep, washing and mending. No senior officer could have relished issuing Colonel-General Curt Haase's rather shamefaced apology to 302 ID the following day. Also, on 18/19 August, 10 Pz was engaged on a map exercise to repel a landing on the coast north of the Somme; and the divisional commander of SS Adolf Hitler and members of his staff were enjoying an overnight visit to the U-boat officers' mess at La Baule.[18] Thus, while the Germans may indeed have picked up the hint of a threat to the Dieppe–Fécamp sector in August 1942 and while there was always the chance of valuable information leaving the country in the diplomatic pouches of neutral embassies, Peis, Farago and Mosely have still failed the test facing all historians of Intelligence, namely, the connecting of source with user. It is not an acceptable method of historical proof to use leapfrogging terms like 'upshot' to explain how a particular piece of intelligence triggered a military catastrophe.

Supposing, though, that the report of 13 August did indeed come from a V-Mann in the south of England, the most believable alternative explanation is surely that there had been a mistake, that the XX Committee had inadvertently cleared accurate information to be passed on to the enemy which luckily was not acted upon. After all, the report of 29 October was not too different from Dragonfly's of 18 October, which also mentioned a possible repetition of the Dieppe raid in conveying the threat posed by Overthrow. The unique process whereby the Dieppe raid was planned, mounted and launched tends to lend this theory a certain persuasiveness. Plans for operations that were cancelled had a built-in plausibility that the deception planners in the Middle East (A Force) exploited more than once. Rutter was cancelled on 7 July and COHQ files reveal that the Examination Committee was prepared to leak the operation two days later as cover for raids being contemplated for September. In addition, CCO did not inform the JIC or ISSB of the resurrection of Rutter and there was no representative from COHQ on the XX Committee. Montagu found attendance at the ISSB meetings useful because it gave him 'assistance [concerning future operations] which my other Service colleagues on the Twenty Committee lacked'; and the MI 5 (Ops) representative on the ISSB relied on its daily meetings to advise whether a proposed operation clashed with plans by 'other authorities' and to take the necessary steps through 'special means' to assist in the execution of other plans.[19] The deception authorities in London, in other words, had no inkling that Rutter had not remained cancelled. Moreover, a COHQ member was appointed to the XX Committee before the end of August in what might be interpreted as damage control. There even seems to have been a precedent of sorts for some such possible misunderstanding, inasmuch as Colonel Oliver Stanley cited an instance at the end of 1941 when the COS changed their minds about a certain operation and switched to another about which he, as Controlling Officer, had been putting out false rumours to cover the first.[20]

On the other hand, some sort of preliminary inquiry into a possible lapse of co-ordination in the use of 'special means' would presumably have been required if this had been the reason for the fairly broad changes made after Jubilee. Ewen Montagu would certainly have heard of it, but he could recall no such post-mortem other than the Sylvan Flakes investigation. The fact nevertheless remains that Jubilee was the cause of a definite rift between COHQ and the ISSB, which reached the COS level. This subject was touched on by Hinsley but without going into the underlying reasons.[21] What were they?

The most useful of the meagre collection of sources available are the Agenda and Minutes of the ISSB. The outline of the dispute is quite clear. The Board had been fully involved in the planning of Rutter, giving expert advice on such matters as the decision to postpone it on 15 June. Troops detailed for the raid were to be led to believe that there had not been a postponement and that they were continuing to train for a large-scale invasion. On 19 August, however, members of the Board were taken aback when a representative from COHQ appeared before them to explain why they had been kept in the dark about that day's raid. From then on the ISSB was at the centre of the post-Jubilee reverberations. On 31 August, for instance, the service members requested copies of the Operation Orders to enable them to study the implications of their capture by the enemy; and on 2 November the results of MI 5's inquiry into the Sylvan Flakes case were reported, with the Board concluding that no further action was called for. From the security point of view, Jubilee was a mess – ironically enough, considering the inordinate but erratic measures COHQ had taken to protect it. Those immediately concerned in Naval Intelligence at the Admiralty (NID) did not get instructions until two days before, whereas 'C' and SOE were informed immediately; these were secret organizations 'but not nearly so secure', according to Commander Ian Fleming RNVR of the NID, 'concerning operations which are not vital to them as are the service departments'. From the ISSB's point of view, the most sensitive issue was procedural, revolving around the Board's terms of reference and COHQ's defiance of them – above all, the principle that the chairman of the ISSB was, by special arrangement, entitled to have prior notification of all operations under consideration. On 18 September, accordingly, the Board resolved that Colonel R. Neville (COHQ) should be made 'fully aware of the Board's procedures regarding operations'. The Board had the support of the JIC, and on 26 December the secretary no doubt had the satisfaction of recording that the COS had invited the three service ministries and COHQ 'to ensure that ISSB was put in the know concerning all operations at the earliest stage'.[22] The ISSB charter was revised to that effect.

More to the point where the underlying reasons are concerned, not even a close reading between the lines supports the interpretation that the issue dividing COHQ and ISSB was at all related to the employment of 'special means'. The crux of the matter was instead the unauthorized appropriation of codenames. The ISSB was responsible for issuing codenames to operational authorities from the Interservice Codeword Index, not a function to be taken lightly. Of the approximately 10,000 codewords used during the war, many were used more than once and the same operation was sometimes known by more than one codename as it went through successive stages of planning and postponement. CCO had already had to be reminded in June that his staff had not followed the correct procedure in the case of Operation Dryad – which in September took shape as a raid on Casquets lighthouse on Alderney – resulting in confusion. But the same staff were inevitably obliged once again to circumvent the ISSB in putting 'Jubilee' into use. Shortly before 19 August, an officer from COHQ had checked all codenames issued by ISSB to that HQ 'for the purpose of the historical record'. He was denied information in cases where words referred to operations that had not taken place but might still be operative, and where the target and operation had political implications. On 18 August, it came to the Board's attention that an indirect attempt had been made by COHQ to obtain the meaning of certain codenames from MO 1, whereupon the Board reaffirmed that such information should still be withheld.[23] Small wonder, then, that its members reacted so strongly the following day.

Returning to the mysterious report of 13 August, a third possibility remains. There is always the chance that the V-Mann was not in the south of England at all but merely passing on a report purporting to have come from a sub-agent there. The wording of other entries in the Naval Staff War Diary proves that the text can support such a reading. On 13 November, for instance, an agent pointed out that a large, heavily escorted troop convoy had left British west coast ports on 11 or 12 November for North Africa, leaving only three King George V class battleships in home waters. Since transatlantic convoys were now being run with greatly reduced escorts, there was a good opportunity for using heavy units of the German fleet against them. This message was described as an 'Ostro-bericht aus England', although the Abwehr were well aware that Ostro himself was based in Lisbon. Unknown to them, Ostro (a Czech named Paul Fidrmuc) was the most gifted of the inventive agents, boasting imaginary sources as far afield as Egypt and Canada, including four or five agents in the UK communicating by secret writing. Ostro's standing with the German High Command rose to unprecedented heights towards the end of 1942 and remained at that level into 1943 and beyond. One of his reports (KTB 1 Skl, 3 October) apparently based on a number of reports (Abwehrmeldungen) from sub-agents, was a masterpiece of invention. Preparations were supposedly under way in the south of England for a major operation. No 11 Fighter Group in the south-east had been withdrawn from operations. Bombers had been moved to stations in the south-west, including Exeter, and delivery flights of new aircraft for the Middle East had been suspended since about 20 September. Terror raids on Germany had been stopped for the previous ten days or so for reasons other than the reputed one of bad weather. Liberators, Whitleys and B-17s were being converted into transports as a matter of urgency. Large formations of tanks had been seen at night on roads north and north-west of Southampton. All leave had been cancelled, arterial roads were clogged with columns of troops and supply traffic, and the main railway stations were like bedlam. Up to 200,000 BRT of shipping had been collected by stopping the return of vessels to the US and Canada since 17 September. Admittedly, this tonnage was not sufficient for a Second Front and some of the preparations such as laying down supply dumps near various ports could more plausibly be intended for an invasion next year. Still, something was in the works, giving a strong impression of 'einer grösseren Aktion'.[24]

'SPY STUFF' – AGENTS, COVER AND DECEPTION

This particular confection circulated through the Intelligence services on both sides like the proverbial barium meal. Von Rundstedt referred to it in his weekly appreciation on 5 October; Foreign Armies West, true to form, labelled reports of tanks north of Southampton as 'unconfirmed' on 10 October. Intercepted and deciphered by GC&CS, the report was in the hands of the CAS by 10 October – one of the rare examples of ISK material to be found in documents so far released at the PRO.[25] It reinforces the point that Allied deception had a way of picking up an echo from inventing agents like Ostro. Ostro was merely supplying his version of the large-scale movement exercise that was the obvious way to add substance to Overthrow. Such a scheme had indeed been planned (Operation Cavendish), although never actually carried out. Ostro was always careful to hedge his bets and astutely blamed earlier misinformation on British cover and deception. On 3 December (KTB 1 Skl), he was quoted to the effect that the apparent build-up for a major operation against the French Channel coast had been partly designed to cover the movement of US troops from Northern Ireland to England prior to their embarkation for Torch. In August 1943 Ostro devised Operation Black Prince, bearing an uncanny resemblance to Starkey, an elaborate British cross-Channel feint culminating on 9 September. Not surprisingly, this rogue elephant of an agent was considered a menace in some quarters of Whitehall.[26]

Ostro was certainly active well before August 1942; according to the Official History, 37 of his reports were intercepted between January and October of that year. A long report at the end of April, when he was still being referred to as an anonymous V-Mann, first mentioned one of his favourite ideas, namely, that it had been a mistake to move the battlecruisers from Brest. He almost certainly contributed to the reports in May about a possible British landing somewhere in western Europe. At least one reference has survived to a 'CHB-Meldung' about a British landing in Norway to take place on 23 May; David Kahn identified CHB 1, CHB 2 etc as members of Ostro's worldwide network of agents. Nigel West mistakenly considered it 'a strong likelihood' that Ostro was the author of A 3924's report in the Naval Staff War Diary of 4 August.[27] Before November 1942 no Abwehr agents had their reports personalized in this style, even the report in the entry for 3 October being simply described as having come from a reliable source. The first mention of Ostro by name in the War Diary was in the report already cited on 13 November, and on 18 November one of his reports was described as the work of an 'extraordinarily well versed and proven' agent. By December this practice had become commonplace, if never quite standardized, even in the spelling of 'Ostro'. Variations ran to 'Abwehrbericht (Ostro)', 'Ostrowbericht', 'Abwehrbericht über Ostrow aus England', and so on. He was touted by the Abwehr as having outstanding contacts, especially in England, and the Naval Staff were quick to point out on 7 January 1943 that one of his reports about rationing on 21 December had just been confirmed in a speech by the Minister of Food. Ostro was still the only agent being singled out in this way and his growing repute seems to have been the result of a cumulative process rather than of any particular scoop. In December 1942 there was a sharp, top-level exchange between the Naval Staff and the Abwehr when the former discovered that they were not on the distribution list for all of Ostro's strategic predictions and recommendations. Many of those recommendations were in favour of stepping up the naval and air war against the British Isles and were being sent directly to Warlimont and the OKW (Operations) staff.[28] Agent Ostro was in a class by himself.

There is a strong possibility that the 'extremely reliable and proven' agent behind the report in the entry for 31 October was Ostro. This was a singular enough characterization and should perhaps be considered the next best accolade to being named. The mention of Dieppe in the text may explain why this credit was cited then rather than in the preamble to the report on 3 October. Predictions of a repetition of the Dieppe raid on a larger scale were quite common in reports from all sources during this period – prompted perhaps by an Abwehr questionnaire – though not all were taken seriously. Only access to a wider selection of sources (assuming they have survived) will reveal what was reported on 13 August and why it failed to make more of an impression until later.

Controlled agents' reports were only one means of channelling cover and deception to the enemy. Cover has been defined as planned measures to disguise or conceal an operation's true objective. All major operations planned in 1942, such as Sledgehammer, to capture a permanent foothold on the Continent, were covered as a matter of course, in the sense that measures were included to achieve surprise and to influence the movement of German reserves. Even the feeblest version of Sledgehammer, to capture Le Havre in the event of a collapse of German morale, came complete with its own cover plan. Small-scale raids, on the other hand, were too frequent and insignificant to require any cover at all, leaving raids on a larger scale somewhere in between. The ISSB decided in March that there were so many operations under consideration against the north coast of France that it was impossible to provide cover for them all. This decision was taken with reference to Chariot and Myrmidon (an operation against Bayonne early in April that had to be aborted because of heavy surf), which were by no means small-scale raids. According to GHQ Home Forces 'Instructions' on 4 June, a medium-sized force preparing to engage on a raid on the Continent could not hope to achieve strategic surprise but should try to preserve tactical surprise by concealing the objective of the raid, the date of launching it, the strength of the force involved and the tactical method to be used. If this seemed feasible, a cover plan would not normally be adopted. If a 'special' cover plan was found to be necessary, however, preparations should be arranged to be visible to enemy reconnaissance and every effort made to ensure that the final assembly of craft and shipping resembled as closely as possible earlier assemblies for exercises. The cover plan should be treated as part of the real plan and should diverge from it at the last possible moment. Lastly, the troops should be shown *Next of Kin* and given the impression that they were engaged on invasion training.[29]

In the case of Rutter and Jubilee, which could only be described as a 'major raid', the decision was made to rely on security. Hughes-Hallett, for instance, could see no point in drawing the Germans' attention to specific activities that they might not have noticed in the first place, since it might not be easy to deceive them about their true purpose. He, of course, was thinking of some sort of visual deception for Jubilee, not being indoctrinated for 'special means'. When Masterman regretted that Jubilee had not been properly covered, he meant by special means – no use had been made of the XX agents to provide the Germans with a misleading explanation for the general preparations on the south coast. Cover for Rutter and Jubilee was in fact restricted to 'defensive' cover, which did not require a direct line of communications to the enemy. Mountbatten had already informed the ISSB that no cover plan was required for Rutter. In Jubilee's case, the ISSB was not consulted at all, beyond being asked on 11 July to judge what risk might be involved in remounting Rutter. While the Jubilee planners were at work, the ISSB was still considering aspects of Rutter's security – 'certain railway movements' and the hospitalization of personnel who had had an accident after having been briefed. The three Force Commanders were interviewed separately in the last week of July and invited to invent homemade cover plans for their own officers and men and such members of the general public as they might come into contact with. Hence curious civilians were told that the combined signals unit billeted in a large house near Aldershot were survivors from a torpedoed troopship; and planning for Jubilee was disguised as staff work on a pamphlet covering experience gained by 2 Canadian division in recent training on the Isle of Wight.[30] Thus, if any leakage *did* occur it would not give away anything of value to the enemy.

The suggestion that cover for Jubilee was wilfully withheld because Churchill and the COS did not want the Germans to be distracted from the operation is the product of Anthony Cave Brown's conspiratorial fantasies. Still, his claim that German 'expectancy' of a raid on Dieppe was aroused by a small-scale raid four nights before Jubilee at least has an interesting documentary context. Minor raids did of course run the risk of keeping the coast in a state of alert, possibly prejudicing the chances of more ambitious operations. Mountbatten conceded this point early in May but argued that minor raids (by anything up to 200 men) would wear the Germans down by forcing them into a semi-permanent stand-to; besides, it was inevitable that they would take extra defensive precautions anyway once the GAF spotted the preparations for larger raids. What he could hardly have anticipated, naturally, was Cave Brown's insinuation that minor raids were intended to provide the enemy with some sort of preliminary warning of larger raids to come, as in the case of the raid on the Cherbourg peninsula on 14/15 August. This was in fact Barricade, a very small raid indeed by only eleven men

'SPY STUFF' – AGENTS, COVER AND DECEPTION

who crossed the Channel in an MTB and came ashore by rubber dinghy. There was the usual confused exchange of gunfire and grenades in the dark, leaving both sides claiming victory. But Barricade had nothing to do with Jubilee and had originally been scheduled for the third week of July in conjunction with Backchat.[31]

Cave Brown may have been misled in this instance by the juxtaposition of the captured documents he used as sources. In February 1944, the Theatre Document Section of SHAEF Intelligence circulated (TDS/SHAEF/86/Feb 1944) six German documents with translations, the two most important of which were LXXXI Corps Ia and Ic reports on Jubilee; two of the others were commentaries on those by LXXXIV Corps and 320 ID. Also included from the files of 320 ID were a divisional order concerning coastal defence of 18 September and a report of 17 August on the landing by rubber boat on 14/15 August in the division's sector south of Barfleur. While it would be correct to say that these documents, together with von Rundstedt's Battle Report which appears to have been captured separately, give a good account of some of the lessons the Germans drew from Jubilee, their texts contain no internal or any other kind of evidence to suggest that Barricade had the kind of admonitory effect (whether intentional or not) that Cave Brown claimed for it on 302 ID or LXXXI Corps. Quite frankly, there is no reason to question the sincerity of Churchill's comment to a COHQ representative shortly after Bristle on 3/4 June: 'I hope those small raids of yours will not have the effect of putting the whole coastline in a state of vigilance.'[32]

If the XX agents were not being used to provide cover for cross-Channel operations in the summer of 1942, what role, if any, were they playing? How were they responding to Abwehr questionnaires about the preparations under way on the south coast of England? The Official History of Deception reveals that there was a division of opinion throughout the summer between those in favour of a bolder policy and operational authorities opposed to providing the Germans with the consequent increase in accurate information. Ewen Montagu and Colonel Robertson of section B1a considered that the time had come to be more ambitious in using the double-cross system to mislead the enemy, while GHQ Home Forces and the Director General of MI 5 still emphasized security rather than deception. In May, Montagu put up a proposal that the XX Committee might achieve something worthwhile if GHQ Home Forces would single out a 'deception target' in France – only to have it emphatically turned down. This entire question was not settled in favour of a more enterprising use of special means until a meeting in September, just in time for Torch.[33]

German sources, generally speaking, can be counted on to reflect variations in the volume of reports and to allow the outlines of Allied cover and deception operations to show through. Montagu and the XX Committee's frustration in June, July and August was translated into a rather imposing but vaguely defined threat from the south of England. Grand Admiral Carls, C-in-C Naval Group North, put it concisely in an appreciation at the end of July: there was plenty of talk about a Second Front but no recognizable focus or 'Zielrichtung'. On 10 August, von Rundstedt admitted that the reported appearance of 5,000 trained parachutists at Tonbridge in Kent gave pause for thought; a week later, he mentioned several reports of there being even more landing craft on the south coast than had been revealed by aerial reconnaissance. But, he admitted, the enemy had disclosed no particular concentrations that might offer clues to his strategic intentions. So far as individual reports are concerned, there is the usual difficulty of trying to identify XX reports in German sources. One good possibility came by wireless from England (KTB 1 Skl, 27 July), according to which Worthing and 'Little Harpton' had been evacuated to make room for invasion troops; anti-invasion minefields and beach obstacles had been removed. This item was a cut above the 'bowdlerized Baedeker' for which Montagu criticized earlier offerings and, if it was indeed the work of the case officer for one of the controlled agents, it showed a fine feeling for authentic detail considering how regularly the Germans mangled the spelling of English placenames.[34] What was conspicuously missing, though, was a build-up of reports indicating an imminent action or pointing in a particular direction, as had been the case with Hardboiled in May and was to be the case again with Overthrow and Solo One in September and October.

Secondly, the deception angle. There are a number of reasons why formal deception – defined as planned measures to mislead the German High Command by passing on a mixture of true and false information regarding strategic plans, strengths and dispositions – did not play a significant role in Allied strategy in the West in 1942. Such a scheme would have been the work of the Combined Commanders' staff, set up in March with authority to plan cross-Channel operations, including large-scale raids, so as to force Germany 'to continuously employ her air forces in active operations and to cause protracted air fighting in the West in an area advantageous to ourselves in order to reduce German air support available for the Eastern Front as early as possible'. The COS, who were inclined to believe that Hitler was a rational man who took the advice of his generals, had already decided that it was most unlikely that the Germans could be forced to withdraw ground forces from Russia. According to their Directive, the Combined Commanders should include in their plans 'a major deception plan' to threaten the enemy with a return to the Continent. But on 14 April, they lost no time in rejecting this recommendation on the grounds that a deception plan on its own would be unlikely to provoke sustained air fighting. Deception was described as a 'passive' approach whose methods had yet to be studied beyond the stage of simply gathering craft and shipping on the south coast. The transfer of 900 dumb barges from the Thames to the south coast, it was added rather lamely, would 'in itself provide a measure of deception'.[35]

Developments on the south coast went beyond the movement of barges, all of which had arrived in the Portsmouth and Plymouth Commands by 1 July (Operation Consular). In March, following a suggestion from CCO, the COS agreed to reopen and expand port facilities, particularly between Newhaven and Poole. The entire coast from the Wash to the Bristol Channel had of course been organized for defence against invasion ever since 1940; now beach obstacles and minefields (like those at Worthing) were to be removed and tugs, lighters, cranes and dockers returned to transfer Southampton and smaller ports into a springboard for a major operation. Only by doing the work well in advance could the security of invasion preparations be protected once the decision was taken to mount it, given that GAF reconnaissance was bound to uncover such activities as the building of hards for the loading of tanks and other vehicles into landing craft, the excavation of an underground HQ at

Portsmouth, and so on. On 19 May, CE and CW convoys between the Thames and the south coast started to run on a two-day cycle, mainly to build up stocks of coal at power stations during the summer period. In addition, the new Combined Operations training and operations base to be opened in May on the Isle of Wight would mean that significant numbers of landing craft, dumb barges and motor boats would be visible for the first time in and around the Solent.[36] The sheer novelty of this display was expected to have an impact on the Germans, as indeed it did, but it was never part of a carefully orchestrated notional build-up and also fell short of what actual invasion preparations would have looked like.

On 21 May, General Wavell urged on the Prime Minister and COS the importance of 'coherent and long-term' planning of deception and the value of 'deceiving and disturbing the enemy by false information'. The COS replied rather defensively that they were well aware of the value of deception but so far had been hampered by lack of resources 'to simulate intentions that were plausible', as well as by the fact that the enemy enjoyed the strategic initiative.[37] They might have added the absence of anything resembling an adequate staff in London to plan and co-ordinate cover and deception. Wavell's telegram coincided with the nomination of Colonel J. Bevan to replace Stanley as Controlling Officer, supported by a meagre staff consisting of 'one Major recently acquired from Western Command and a Flight Lieutenant RAFVR who had hitherto acted as "Secretary" to the Controlling Officer'. The Flight Lieutenant was the novelist Dennis Wheatley, who had almost single-handedly pushed Hard-boiled through against considerable bureaucratic and administrative opposition. On 3 June, Bevan stated that the object of deception in the coming months should be the creation of uncertainty in the mind of the enemy about British strategic intentions, without getting down to details about the timing and location of possible operations. He briefly worked on a cover plan for Sledgehammer before it was cancelled and then directed his attention to Torch, presenting his cover plans on 5 August.[38] But for most of the summer it was easier to create uncertainty in the mind of the enemy than to bring it under control at home.

In the spring of 1942, the COS faced a number of contingencies, ranging from a 'desperate venture' for political rather than military reasons should the Russians be on the verge of defeat, to an even more unlikely operation to take advantage of a sudden collapse of German morale such as had occurred after August 1918. This uncertainty was bound to have a serious effect on training, as C-in-C Home Forces pointed out on 31 August. Troops under his command had successively to prepare for 'a minor operation' in 1942, 'a major operation' in 1943 in conjunction with the Americans, and again for 'a lesser operation' in 1942, plus providing units for training for possible operations in the Arctic or mountainous country, finding large drafts for the Middle East, and of course ensuring the defence of the British Isles. For each project a different kind of army was required and six months were necessary to train the troops in their new role. The result, according to Paget, had been 'disorganization and waste of effort'.[39]

Quite apart from that, however, there was the threat in the spring of conflict between notional and what might turn out to be actual operations, not to mention the realization that a mere cross-Channel feint could never satisfy the growing public clamour for a Second Front or Russian and American expectations of cross-Channel operations. Stanley indeed had observed in March that a major deception that year could have 'an adverse effect' on public opinion. Having ruled out deception, the Combined Commanders faced a choice between seizing a permanent bridgehead on the Continent (Sledgehammer) or relying on raiding as the answer to doing something for the Russians should they be hard pressed that summer. It was soon found that a permanent bridgehead within the advantageous zone – corresponding to the Spitfire's operational range, that is, between Dunkirk and the Somme estuary – would have to be expanded to include a deepwater port because Calais and Boulogne could be easily blocked and in any case lacked the capacity to support such an expedition. But the resources to capture Antwerp or Le Havre were simply not available, and such an operation was not feasible at all outside the advantageous zone. A major raid would be unlikely to work by itself but a series of raids 'and the threat of more to come' might force the Germans to withdraw air forces from other theatres to reinforce those on the Channel coast. On 14 April, accordingly, the COS endorsed a series of medium-sized raids as 'the only practicable solution'.[40]

The proposal to seize a permanent bridgehead in France, unfortunately, was not to be so easily disposed of, because the COS recommended that the Combined Commanders continue to study it so as to be in the position to take advantage of an unexpected opportunity. By May, the Combined Commanders had reached the decision that Sledgehammer would be a feasible operation only if German morale collapsed; and by 27 May the Prime Minister had accepted the view that such an operation against unimpaired German opposition would be purely sacrificial and on too small a scale to help the Russians anyway. On 8 June, he ruled that there should be no substantial landing in France except in the event of German demoralization and unless the intention was to remain. Sledgehammer now lost its connection with provoking an advantageous air battle and became increasingly associated with deception, mainly directed against the Germans but also towards the Russians, who were counting on such an operation being carried out.[41] By mid-May, in fact, the outline of a tidy solution to the dilemmas of British strategy in 1942 could be discerned: a series of raids on an ascending scale of intensity, combined with a gradual build-up on the south coast for a notional Sledgehammer in September. In the event, such were the pressures and constraints bearing on the Prime Minister and COS during the grimmest months of the war, that not only did this model fail to come together but its constituent parts fell apart as well.

Sledgehammer died a lingering death. On 11 June, the War Cabinet recommended that the necessary shipping and landing craft be taken up for the sake of exerting strategic pressure across the Channel. But Mountbatten had already investigated the implications of withdrawing landing craft from training for Roundup, the cross-Channel invasion in 1943, and found that an unacceptable postponement would be the result. The Ministry of War Transport complained that the loss of 200 coasters from service would saddle the already over-burdened railways with a further 500,000 tons, mostly of coal, and the diversion of other shipping would cut imports by about 250,000 tons a month.[42] The COS were torn between openly declaring Sledgehammer to be impracticable, in which case they feared for the effect on Russian morale, or paying a considerable material price for the sake of bluff. At the same time, the US Joint Chiefs persisted in trying to breathe life into their version of Sledgehammer designed to capture the Cotentin peninsula later in 1942. That idea was finally put to rest at the meetings with General Marshall in London, on 22–24 July, which had far-reaching and irreversible consequences for the rest of the war in the shape of the decision to go ahead with the landings in French North Africa; but on 22 August the CAS found it necessary to remind his fellow Chiefs of Staff that they were still committed to continuing preparations for Sledgehammer for the sake of deception, although without specifying what form those preparations might take.[43] The operation was finally put out of its misery by the adoption of the cover plans for Torch.

The series of medium-sized raids never came to anything either. Mountbatten spoke confidently at the end of April of launching a sizeable raid once a month or so, with small-scale raids interspersed roughly every two weeks; on 22 May he promised two Soviet representatives to carry out such a programme 'irrespective of the situation'. Hughes-Hallett, indeed, had drawn up a schedule that Mountbatten had accepted in principle on 23 January. After Bruneval in February, there was to be a descent on a house on the sea-front at Ostend which the GAF used as a convalescent home for pilots, followed by St Nazaire at the end of the month and Myrmidon (Bayonne) early in April. The May raid would be an attempt to capture Alderney (Blazing) and hold it as long as possible so as to disrupt the convoy traffic with the U-boat bases to the west. In June and July Dieppe was to be the target for the double raid already mentioned, which was to be in divisional strength. In August, a bridgehead south of the Somme would be seized from which a force of armoured cars could make a dash on Paris.[44]

The planning of these and other raids took place in a highly charged atmosphere which grew in intensity as one cancellation or misfire followed another. Myrmidon was a flop and Blazing was cancelled on 6 May. Plans under consideration by the Examination Committee at COHQ had a way of appearing under one code-name, disappearing, and then reappearing, amoeba-like, under another. Barricade did not change its name but in its July version was large enough to require the participation of an LSI and was aimed against a suspected coast-watching radar at Barfleur. Aimwell was to be Blazing's successor, to capture

Alderney and use it as a base for an operation against Cherbourg (Gabriel). While Gabriel was in progress, Myrmidon would be tried again to take advantage of the preoccupation of the GAF in the Cherbourg area. Bristle, one of the few operations to get off the drawing board, had its origin in five simultaneous raids north and south of Boulogne, which were then reduced to three (Lancing) when pressure was felt to lay on something quickly after the cancellation of Blazing. Bristle, which was Lancing reduced to one raid too small to qualify as medium-sized, took place south of Boulogne on 3/4 June. Some of the planners, according to a senior member of the COHQ Executive, were guilty in their chopping and changing of '. . . chronic individualism amounting almost to indiscipline'.[45]

The Combined Commanders had even less success than COHQ. The large-scale raid conceived by their planning staff (Imperator) in response to a directive from the COS looked more like a 'desperate venture' than a serious act of war, having originated as a raid on Paris by an armoured force of 60 tanks. As an additional complication, re-embarkation was to be through a different beachhead from the one captured in the initial assault. Imperator, charitably described as 'bold in the extreme', was apparently taken seriously, with a target date in July or August, but it was cancelled on 8 June. In May, the COS and Combined Commanders felt themselves under mounting pressure 'to do something on the Continent', as Leigh-Mallory put it on 25 May. His suggestion – 'A Suggested Alternative to Sledgehammer: A New Alternative to Imperator' – was to assault and capture Boulogne and the high ground surrounding it, transform it into a fortress with strong landward and air defences, and hold it permanently. Such an action would go far towards meeting the 'persistently increasing' demands of the British public and the Russians and Americans. Nevertheless May passed without a medium-sized raid, as did June, after the postponement of Rutter; with its cancellation on 7 July, there was every chance that July and August would witness no more impressive offensive activity than that provided by the likes of Backchat and Barricade. Understandably enough, it was in a mood of semi-crisis that the Examination Committee at COHQ met on 9 July to decide what was to take the place of 'abandoned [sic] Sledgehammer and dismounted Rutter'.[46]

'SPY STUFF' – AGENTS, COVER AND DECEPTION

The Committee proposed three raids, confusingly codenamed Sledgehammer East, Central and West, against Boulogne, Alderney and St Nazaire respectively, to be launched early in September. Actual preparations for the raids, plus additional measures such as loading exercises with barges and the stationing of MT ships in the Thames estuary, would serve the needs of the Committee's second assignment, namely, a deception scheme to simulate preparations for Sledgehammer. Two divisions would be briefed for attacks on Le Havre and Dieppe later in September. Rutter, interestingly enough, could now be leaked and Rutter maps distributed to the masters of the MT ships. This scheme was killed by the Torch decision on 22 July, if not before, by the highly secret decision on 11 July to remount the raid on Dieppe, and no action was taken on the cover and deception proposals. There was clearly a strong likelihood of a triple disaster: Boulogne, to be stormed in a frontal assault, was an important staging port for the coastal convoys and more heavily defended than Dieppe; Alderney's tides presented serious navigational difficulties and the Germans had laboured all summer to turn the island into an impregnable fortress; St Nazaire was far beyond the range of fighter cover and the weaknesses in its defences exposed by Chariot had been corrected. It is less than reassuring to read in the COHQ War Diary that the success of any one of the raids would pay dividends even if the other two failed.[47] Given that the growing strength of Le Havre's and Cherbourg's defences was fast excluding them from consideration as targets for even the heaviest of raids, it is reasonable to ask what worthwhile but still operationally feasible targets for medium-sized raids were left on the Channel coast.

Compared with the three Sledgehammers and Imperator, Jubilee was a model of military prudence. Perhaps it is just as well that, for all the planning activity at COHQ, there was no ready-made plan or force trained for another major or medium-sized raid or raids to replace 'dismounted Rutter'. Above all, though, Jubilee was a fitting climax to a summer of instability and disarray. Instead of a series of raids, there was only the one – set in the volatile context of a 'war of nerves'. To the degree that it was under official control at all, this war was waged by the Political Warfare Executive (PWE), which based its campaign in 1942 on two main propositions: that there would be a Second Front somewhere in the West that summer, and that the Germans would fail to win a decisive victory in Russia. PWE certainly pushed the Second Front theme for all it was worth through rumours spread in diplomatic circles abroad. The ambassador to Portugal, for instance, was reported by his German counterpart as having admitted on 7 July that earlier talk about a cross-Channel invasion had been for deception purposes; but public opinion had become so inflamed that the government was now under pressure to follow through on its promises. On 6 August, there was great excitement in Lisbon over a rumour from London, which may or may not have been the work of PWE, according to which Churchill had been presented with an ultimatum from Stalin to launch a Second Front within the next twelve days or the Russians would no longer consider themselves bound by the terms of the Anglo-Soviet Treaty signed in May.[48]

Against this background, Jubilee stood out for what it was: an improvised exploit. There was no time for PWE to assimilate the operation into some purported General Plan for Victory. A series of raids might have lent plausibility to the promotion of a Second Front but not an operation on its own that the Germans could deride as a failed invasion. Any raid while the current situation continued in Russia, according to Peter Murphy, a senior PWE official, in a paper on the diminishing value of combined operations on 26 July, could have 'disastrous political repercussions'. He had already pointed out on 27 May that 'a slump in the tonic value' of raids for the British public was soon to be expected. As the novelty wore off and hopes of bigger and better things to come remained unfulfilled, a certain amount of sales resistance would set in. By the end of July, according to Murphy, the law of diminishing returns had produced just such a reaction against a background of uncontrolled agitation for a Second Front in the Soviet Union, the United States and at home. In short, failure to integrate Jubilee into a larger strategic plan for publicity and propaganda purposes left the impression of a sporadic and haphazard policy. Not surprisingly, the British lost the propaganda battle over Jubilee, as they had earlier in the year over Chariot. It was particularly regrettable, of course, that news of Churchill's visit to Moscow – 'a prisoner of the Kremlin', according to Goebbels – broke on 18 August.[49]

'SPY STUFF' – AGENTS, COVER AND DECEPTION

Underlying the events of the summer of 1942 in the West was a tension between mounting a notional strategic threat by means of a 'war of nerves' and launching actual operations heavily dependent on tactical surprise. Whether Churchill liked it or not, small-scale raids were likely, indeed were intended, to put the coast in a 'state of vigilance'; he just happened to be thinking of the requirements of the forthcoming raid on Dieppe when he made his remark. Possibly the best example of this particular tug-of-war was the PWE-sponsored broadcast by the BBC on 8 June in the form of an appeal-cum-warning to the civilian population of the coastal zone from Belgium to the Pyrenees to evacuate their homes so as to avoid being caught up in impending Allied operations. On 8 June, Rutter's first possible launching date was less than two weeks ahead. As Cave Brown would have it, this warning, repeated nightly thereafter, was an attempt at manipulating the BBC and French listeners for the purposes of deception. Actually, the Foreign Office, COS and COHQ had been considering such a step since the end of April, partly to spare French lives during a preliminary naval and air bombardment and partly to saddle the Germans with the moral responsibility for any unavoidable loss of life. Mountbatten pointed out in May that such a bombardment would be an 'inseparable part' of any future raids.[50] Ironically enough, the heavy preliminary bombing of Dieppe was dropped from the Rutter plan three days before the first broadcast. The COS and Foreign Office were also well aware of the danger of a preliminary uprising by the 'patriot armies' on the Continent, and the broadcast contained a reminder of the threat of retribution for helping Allied troops. On 8 July, the Foreign Secretary told Bruce Lockhart that the evacuation warning should be dampened down because the French were becoming dangerously excited about a Second Front in August. According to the Ic of 302 ID, half the population of Dieppe had taken to sleeping in the surrounding countryside for the rest of the month after 8 June. The Germans themselves took the warning in their stride, at least before the third week of June, as just another incident in the *Nervenkrieg*.[51]

The apparent affront to Rutter's security was rationalized by the COS by claiming that the Germans would have become accustomed to the repeated warnings by the time Rutter was ready to go. Meanwhile, what Churchill called the 'public clamour' on behalf of a Second Front continued unabated, with every public figure from Liddell Hart to H.G. Wells voicing strong views on the subject. 'Official sources' often joined in. On 16 July, for example, all the London dailies ran portentous reports about 'the greatest raid and invasion manoeuvres ever carried out in European waters'. An exercise had just been completed on the south coast by a powerful Canadian force under the eyes of Mountbatten, McNaughton and Crerar, possibly as a prelude to 'what may be military operations on the European coast'. Although not instantly recognizable, this was a Ministry of Information reference to Exercise Yukon II on 22/23 June.[52] It need hardly be pointed out that the solution to the predicament of including actual operations as part of a war of nerves was not Mountbatten's absurd notion of somehow tiring the Germans out: the solution was to take the most extraordinary precautions to preserve tactical surprise. Tactical surprise, moreover, was less a matter of suppressing careless talk, or even catching spies, than of gathering intelligence about the German system of early warning, consisting of aerial reconnaissance, radar, off-shore patrols and other methods of coastal surveillance.

So far as cover and deception were concerned, finally, Jubilee had very little to do with the former and next to nothing with the latter, even in the loose form of psychological warfare. At least the raid did no damage to the XX system, and the threat of a repeat operation was put to some use for Overthrow. That the raid might have been somehow saved by a cover plan must remain doubtful, as will be argued later. Also doubtful is the proposition that the 'tidy' strategic solution of raids building up to a notional Sledgehammer in September would necessarily have produced better results. After all, the 'major deception plan' which the Combined Commanders rejected for 1942 was implemented in 1943 by COSSAC, Lieutenant-General F.E. Morgan, whose primary assignment was to plan the cross-Channel invasion in 1944. Morgan's deception plan, complete with preliminary and preparatory air operations, an amphibious feint and full use of 'special means', produced no reaction at all, except for some emergency mine-laying, whereas in 1942 the Germans did at least reinforce their ground forces in France. This difference should probably be explained by the fact that Germany's resources were more tightly stretched in 1943 than in 1942 rather than by the respective merits of a 'war of nerves' and planned strategic deception. On the other hand, it might be argued that a well-planned deception in 1942 would have produced still better results. That is one of a number of interesting questions only the German documents can answer.

NOTES

1. Abwehrstelle im Wehrkreis X, B Nr 78/43, I H geh, 22.1.43: 'Zuv A 3924 meldet am 19.1.43 aus Paris (Quelle franz Kolonialkreise) . . . (Zusatz Ast X: Meldung wird mit Vorbehalt vorgelegt. V-Mann ist derselbe d Landung Dieppe 14 Tage vorausmeldete.' T 77 1539.
2. Abwehrstelle im Wehrkreise X, B Nr 895/42, I H geh, 4.8.42: 'Zuv Va 3924 meldet brieflich aus Paris am 29.7.42. In franz anglofilen Kreisen herrscht plötzlich grosses Interesse für den Hafen von Le Tréport und Umgebung . . .' T 77 1539.
3. KTB, 1 Skl, Teil A, 15.3.43, T 1022 1679. Abwehrstelle im Wehrkreis X, B Nr 534/43, I M geh, 15.2.43: 'Zuverl und bewährter R 3764 (V-Mann I Wi, der seinerseit auch rechtzeitige Nachricht über Dieppe und Schiffszusammenziehung in Gibraltar vor Afrikalandung brachte) . . .' T 77 1537. KTB, 1 Skl, Teil A, 6.8.42, T 1022 1675.
4. Ast Hamburg, B Nr 273/42, I H geh, 8.11.42: 'Zuverlässiger A 3924 meldet am 8.11.42 aus Paris . . . Angriff auf Casablanca von See her zwischen 13 und 17 November.' T 77 1537. Ast Hamburg, B Nr 1258/42, I H geh, 4.11.42: 'Zuverl R 3917 meldet am 4.11.42 aus Paris . . . Amerikanische Landung in Casablanca innerhalb von vierzehn Tagen.' T 77 1539. Hinsley, *British Intelligence*, Vol. 2, pp. 480, 478. An agent, for example, was given credit for having given forewarning of Husky in KTB der Kommandanten der Seeverteidigung Seine-Somme, 23.7.43, T 1022 2371. Heinz Höhne, *Canaris* (trans J.M. Brownjohn, London, 1979), p. 491. Reile, *Geheime Westfront*, p. 326.
5. The 'recht wenig vertrauenerweckend' is from KTB, Marinegruppenkommando Nord, 14.6.42, RM 35 I/142, concerning two agents' reports of landings in N. Norway and N. Denmark and of 80,000 US troops near Glasgow. 'Lagebeurtailung durch Ob West', Ia Nr 3013/42 gKdos, 5.10.42: 'In Raume um Salisbury erstmale Amerikaner erfasst.' T 78 311. Reile, *Geheime Westfront*, p. 270.

6. *KTB, OKW(WFSt)*, II (1), 19.5.42, p 368: 'Meldung von der *Abt Fremde Heere West des GenStdH* über ein geplantes brit Landungsunternehmen des Vizeadmirals Mountbatten.' The spelling in other sources is still usually 'Mount Batten'. KTB, 1 Skl, Teil A, 19.5.42, 23.5.42, T 1022 1673.
7. Dennis Wheatley, *The Deception Planners: My Secret War* (London, 1980). WO 106/1987, Operation Hardboiled, contains very little information. Howard, *British Intelligence*, Vol. 5, pp. 23–4. KTB, 1 Sk1, Teil A, 30.5.42, T 1022 1673.
8. The reference to ABC is to be found in an 'Abschrift' in the Ic files of LXXI Corps in Norway (T 314 1557). The uncontrolled agent 'Josephine' was occasionally referred to as source 'Alpha/Beta'.
9. KTB, 1 Skl, Teil A, 23.5.42, T 1022 1673. 'Einzelbefehl des Ob West Nr 4,' Ia Nr 1363/42 gKdos, 23.5.42; the 'Abwehr- und Alarm-Bereitschaft' was lifted on 26 May ('Einzelbefehl des Ob West Nr 5,' Ia Nr 1418/42 gKdos, 26.5.42), T 78 317. WO 219/1933, Martian Report No. 8, 21 July 1942.
10. KTB, 302 ID, Ia Anlageheft D zum Kriegstagebuch Nr 3: 'Gefechtsbericht vom 12./13.6.42'; Ob West, Ic/AO Nr 1634/42 g, 26.6.42, T 315 2019. KTB der 2 Sicherungsdivision, 18.7.42, T 1022 3552.
11. KTB des Marinegruppenbefehlshabers West, 21.8.42, T 1022 3974. WO 208/3573, MI 14 Weekly Summaries for CIGS: Summary for week ending 31 August 1942. KTB, 1 Skl, Teil A, 17.10.42, 29.9.42, T 1022 1676. GenStdH, Abt Fremde Heere West III, 'Lagebericht West Nr 734', 31.10.42. Handakten Etzdorf (Vertr AA beim OKH) betreffend Lageberichte West Nr 521–712 vom 10.9.41 bis 19.8.42 (Microfilm purchased from Politisches Archiv des Auswärtigen Amtes, Bonn). Masterman, *Double-Cross System*, p. 111.
12. *KTB, OKW(WFSt)*, II (2), 5.10.42, p. 797 (Erläuterungen des Generals Warlimont). OKW/WFSt/Op(H), Nr 003667/42 gKdos, 9.10.42, T 78 317. KTB, 302 ID, Ia Anlageheft C zum Kriegstagebuch Nr 3: Armeeoberkommando 15, Armeebefehl!, Ia Nr 5770/42 gKdos, 5.10.42, T 315 2019.
13. KTB, 1 Skl, Teil A, 15.3.43, 15.2.43, T 1022 1679; KTB, 1 Skl, Teil A, 31.10.42, T 1022 1677: 'Nach Abwehrbericht aus Südengland vom 29/9 soll für den kommenden Monat eine ähnliche Unternehmung wie die Aktion Dieppe zu erwarten sei. Abwehr berichtet dazu, dass gleicher, ausserst zuverlässig V-Mann, auf

den vorstehender Meldung zurückgeht, auch unter den 13.8 das Unternehmen Dieppe vorausgemeldet hat.'
14. Günther Peis, *The Mirror of Deception: How Britain turned the Nazi spy machine against itself* (trans William Steedman from *So ging Deutschland in die Falle* (1976), London, 1977), pp. 120 ff. *Sunday Times*, 25 Jan. 1981: '"How I cheated Hitler" by the man sent by the Führer to spy on Britain.'
15. Ewen Montagu, letter to the author, 13 Oct. 1982. Masterman, *Double-Cross System*, pp. 127, 108. Nigel West, *Unreliable Witness, Espionage Myths of the Second World War* (London, 1984), p. 96; Richard Deacon with Nigel West, *Spy!* (London, 1980), p. 123.
16. Ladislas Farago, *The Game of the Foxes: The Untold Story of German Espionage in the United States and Great Britain During World War II* (London, 1971); *Daily Express*, October 1971. The series began with an introduction by the editor, Peter Grosvenor, on 16 Oct., entitled, 'The Fantastic Farago'; the Dieppe instalment appeared on 25 Oct. Hinsley, *British Intelligence*, Vol. 2, p. 20.
17. Leonard Mosley, *The Druid* (London, 1982); West, *Unreliable Witness*, pp. 91–4; Phillip Knightley, 'Dinner with the Spymaster', *Sunday Times*, 15 March 1981; Howard, *British Intelligence*, Vol. 5, p. 20.
18. KTB B, AOK 15, Ia Einzelbefehle d Armee: 'Fahrten durch das Armeegebiet', Nr 51/42, 18.8.42; Armeebefehl, 19.8.42: 'Er [the enemy] hat damit das getan, was ich in meinem Befehl vom 18.8 Nr 51/42, Ziffer 4, als *ungewoehnlich* bezeichnet habe.' T 312 502. RG 331, Box 32, Martian Reports Nos 29–78, Martian Report No. 71, 13 Oct. 1943, captured document circulated by GSI(a), HQ 21 Army Group (ré 10 Panzer); KTB, SS Adolf Hitler, 18.8.42, T 354 611.
19. DEFE 2/2, War Diary, COHQ, Examination Committee Meeting, 9 July 1942; Hinsley, *British Intelligence*, Vol. 2, Appendix 13: 'Intelligence Before and During the Dieppe Raid', p. 697. Ewen Montagu, *Beyond Top Secret Ultra* (London, 1978), p. 105; WO 283/6, ISSB Agenda and Minutes, 31 March 1942.
20. Masterman, *Double-Cross System*, p. 109. FO 898/24, Minutes of Executive Meeting (PWE), 18 Nov. 1941. According to Ralph Bennett, GC&CS was not informed in advance about operational plans at this stage of the war (*Ultra and Mediterranean Strategy*, p. 187).
21. Ewen Montagu, letter to the author, 13 Oct. 1982: 'If my word is

not enough I can say that there was an enquiry about possible leaks – one was an advertisement in a newspaper (*The Daily Telegraph*?) which mentioned "a holiday soon in Dieppe" (or some such words) just before the raid. It turned out that it had been sent to the paper long before the raid was thought of and merely came then to the top of the queue waiting for space owing to newsprint rationing. That enquiry would not have happened in that form if we had, in fact, inadvertently sent such a message.'

22. WO 283/7, ISSB Agenda and Minutes, 15 June, 24 June, 27 June, 19 Aug., 31 Aug., 2 Nov. 1942. On 24 June, Bevan proposed that 'certain information' about Rutter should be passed to bomber crews operating over Germany; although the Board considered this an excellent proposal, it was turned down by COHQ and the Air Ministry. DEFE 2/333, Fleming to CCO (attn Casa Maury), 21 Aug. 1942. WO 283/6, ISSB Agenda and Minutes, 18 Sept., 26 Dec. 1942. Hinsley, *British Intelligence*, Vol. 2, Appendix 13: 'Intelligence Before and During the Dieppe Raid,' p. 697.
23. WO 283/7, ISSB Agenda and Minutes, 13 June 1942: 'A Minute to CCO was drafted, pointing out that the proper procedure for informing the Board when a code word was taken into use had not been observed on this occasion [Dryad], and as a result confusion had arisen.' ISSB Agenda and Minutes, 18 Aug. 1942.
24. KTB, 1 Skl, Teil A, 13.11.42, T 1022 1677; 3.10.42, T 1022 1676. Ostro's authorship of the report on 3 October is confirmed in Ausl Abw I, Abw Nr 6452/42 Abw I, 3.10.42, T 1022 2108.
25. 'Lagebeurteilung durch Ob West', Ia Nr 3013/42 gKdos, 5.10.42, T 78 311. AIR 20/4504, AI 3(a)2, 8 Oct 42: 'On 3rd October a report (ISK No. 18825) was forwarded to Berlin . . .' ISK or ISOS stood for Abwehr Enigma decrypts.
26. KTB, 1 Skl, Teil A, 3.12.42, T 1022 1678; KTB, 1 Skl, Teil A, 13.8.43, T 1022 1682.
27. Hinsley and Simkins, *British Intelligence*, Vol. 4, p. 199. Authorship of the report about the battlecruisers is confirmed in KTB, 1 Skl, Teil B, A Ausl/Abw, B Nr 50/43 gKdos I M, 12.1.43, T 1022 1708. MI 14/543, Fernschreiben, OKW/WFSt/Op(H), Nr 01381/42 g, 20.5.42; David Kahn, *Hitler's Spies, German Military Intelligence in World War II* (New York, 1978), p. 356. West, *Unreliable Witness*, pp. 94–5.
28. KTB, 1 Skl, Teil A, 11., 18.11.42, T 1022 1677; KTB, 1 Skl, Teil A, 7.1.43, T 1022 1678. KTB, 1 Skl, Teil B, A Ausl/Abw, B

Nr 50/43 gKdos I M, 12.1.43, T 1022 1708: 'Wunschgemäss werden sie [Berichtsmaterial u Vorschläge Ostros] nunmehr auch der Sk1 zur Kenntnisnahme übersandt.'
29. WO 199/3008, Operation Sledgehammer: Memorandum to COS, CC(42)45(Final), 31 July 1942. WO 283/6, ISSB Agenda and Minutes, 4 March 1942. RG 24, C 17, 10871, 232C2 (D 29) Intelligence: General Security Instructions to Military Force Commanders, GSI(b), HF/Ib/9/37, 4 June 1942.
30. Hughes-Hallett, 'Memoirs', pp. 171–2. Masterman, *Double-Cross System*, p. 108. WO 283/7, ISSB Agenda and Minutes, 11 July-3 Aug. 1942: the Board decided that there would be 'considerable risk' but with that the subject was dropped. DEFE 2/334, HQ 1 Canadian Army, Intelligence Report, 8-5-1 Ops, 56-1-1 Int, 22 Sept. 1942.
31. Anthony Cave Brown, *Bodyguard of Lies* (New York, 1975), p. 81. CAB 80/62, COS(42)130(O), 9 May 1942: Minor Raids: Memorandum by the Chief of Combined Operations. DEFE 2/109, Operation Barricade. Barricade was originally intended to destroy the 'ship locating' RDF station at Barfleur, but the raid as carried out bore little relation to the first plans drawn up in May.
32. This package of documents – 'German Ia and Ic Reports on the Cotentin and Dieppe Raids 14/15 and 19/8/42' – is to be found in a number of places – in DEFE 2/338 at the PRO, under MI 14/830 at the IWM, as well as at the NA (TDS/SHAEF/86, 24 Feb. 1944), where Cave Brown used them. DEFE 2/65, Churchill quoted by Lt. Cdr. de Costobadie in his report on Bristle to CCO, 4 June 1942.
33. Howard, *British Intelligence*, Vol. 5, p. 12.
34. KTB, MarGpN: 'Zusammenfassende Lagebetrachtung Ende Juli', 31.7.42, T 1022 3950. 'Lagebeurteilung durch Ob West', Ia Nr 2436/42 gKdos, 17.8.42, T 78 311. KTB, 1 Skl, Teil A, 27.7.42, T 1022 1675.
35. CAB 79/56, Directive to Combined Commanders – i.e. C-in-C Home Forces, AOC-in-C Fighter Command and CCO – (Annex to COS(42)12th Meeting(O), 21 March 1942). CAB 80/62, Joint Memorandum on Sledgehammer, COS(42)99(O), 14 April 1942, which included a summary of the first report on Sledgehammer, 27 March. 1942.
36. DEFE 2/650A, Operation Consular; CAB 79/56, COS(42)12th Meeting(O), 21 March 1942: discussion of Memorandum by CCO

on preparations for opening south coast ports. ADM 199/2112, Trade Division History, Vol. 34, pp. 369–70.
37. Quoted in full by Howard, *British Intelligence*, Vol. 5, pp. 25–6; original in CAB 120/769, Most Secret 12461/C, 21 May 1942 – 'Private for Prime Minister from General Wavell'. Reply (02 540), 21 June 1942.
38. WO 283/6, ISSB Agenda and Minutes, 3 June 1942; on 7 Aug., the Board resolved that it was a matter of urgency that Bevan should be provided with the necessary staff. AIR 20/4504, 'Cover and Deception'. L.C. Hollis to VCOS, 29 Aug. 1942. Bevan's paper – LCS(42)1 – was discussed at COS(42)84th Meeting(O), 10 Aug. 1942.
39. WO 199/451, Paget to Under-Secretary of State, War Office, HF S/00/199/Ops, 31 Aug. 1942.
40. CAB 79/56, COS(42)65th Meeting(O), 6 July 1942: it was agreed that recommendations should be made as to what if any preparations should be made for the purpose of (i) deceiving the enemy and (ii) of not leading the Russians to assume that there was no chance of an invasion in 1942. CAB 80/62, COS(42)99(O), 14 April 1942. Stanley quoted in the first report on Sledgehammer (P 123/698, 27 March 1942), which was incorporated into the Memorandum of 14 April.
41. CAB 79/56, COS(42)46th Meeting(O), 27 May 1942; COS(42)52nd Meeting(O), 8 June 1942.
42. CAB 79/56, COS(42)61st Meeting(O), 30 June 1942; COS(42)62nd Meeting(O), 1 July 1942.
43. CAB 120/82: Visit of American Chiefs of Staff, July 1942; CAB 79/56, COS(42)97th Meeting(O), 22 Aug. 1942.
44. DEFE 2/2, War Diary COHQ, 22 May 1942; Hughes-Hallett, 'Memoirs', p. 118.
45. See entries in DEFE 2/2, War Diary COHQ, for May and June 1942; DEFE 2/109, Operation Barricade; DEFE 2/911, Meetings of COHQ Executive 1942: Meeting of 16 December – the officer quoted was Brigadier G.E. Wildman-Lushington.
46. DEFE 2/306, Operation Imperator: Imperator was later planned with a more limited objective – see summary by Col A. Head (P 129/1740, 6 June 1942). CAB 106/1027, Combined Commanders' Meetings, 'Operations in France', 11G/TLM, 25 May 1942. DEFE 2/2, War Diary COHQ, 9 July 1942.
47. DEFE 2/2, War Diary COHQ, 9 July 1942; DEFE 2/561, Operation Sledgehammer East.

48. CAB 84/45, Political Warfare against Germany, PW(M)(42)34, 9 July 1942. AKTEN betreffend 'Zweite Front': Lissabon Nr 2186, 7.7.42; Lissabon Nr 2587, 6.8.42, T 120 165.
49. FO 898/375, Peter Murphy to Brig R.A.D. Brooks, 26 July 1942; DEFE 2/546, 'Rutter Notes for Publicity', 29 June 1942, Appendix 6 to Miscellaneous Papers, anticipated many of such developments.
50. FO 898/13, PW(E)(42)47, 1 June 1942: Campaign to the French Coastal Populations. Cave Brown, *Bodyguard of Lies*, p. 83.
51. Kenneth Young (ed.), *The Diaries of Sir Robert Bruce Lockhart* (2 Vols, London, 1980), II, p. 181. KTB, 302 ID, Ic Anlageheft E zum Kriegstagebuch Nr 3: 'Englische Propaganda', Abt Ic 288/42 g, 14.6.42, T 315 2019.
52. CAB 79/56, COS(42)46th Meeting(O), 27 May 1942. The Germans followed the public debate about a Second Front (eg AKTEN betreffend 'Zweite Front': Bern Nr 1113, 7.7.42, T 120 165) but attached no significance to the fact that Liddell Hart was sceptical while H.G. Wells was a strong supporter. *The Times*, *Daily Express*, etc, 16 July 1942.

3

Channel Watch 1942

For German Intelligence, the pivotal event of 1942 in the West was the discovery by 3 Air Fleet that there had been a striking increase in the number of small craft along the English south coast from 1,146 on 3 June to 2,802 on 23 June. Photographic reconnaissance missions had been flown between Dover and Plymouth; on 25 June, coverage was extended as far west as Falmouth. According to the GAF summary included in the Naval Staff War Diary on 25 June, vessels were divided under three headings: 'Harbour and coastal', 'Small vessels up to 15m long', and 'Large vessels and barges'. Landing craft were not accounted for in any systematic way. For a start there was no mention at all of Tank Landing Craft (LCT), two flotillas of which had been on the south coast since the start of June and one since the start of May. Assault Landing Craft (LCA) were located in unspecified numbers for the first time at Chichester on 23 June. Enlargements of photographs taken on 22 June picked out 185 'special' landing craft, whatever that meant, between Portsmouth and Portland. A large number of small specialized craft of a type never seen before were drawn up on shore at the mouth of the 'Exbury' south of Southampton and in Poole harbour. To confuse matters further, there was an increase in boats 20–30 m long in the ports of Southampton and Portland and especially in estuaries and inlets nearby. The Air Fleet concluded that an enemy landing should be considered imminent but did not rule out the possibility of a large-scale feint, since the vessels were grouped closely together more or less in the open.[1]

This reconnaissance had been specially laid on at the request of C-in-C West, following the surge of rumours after Molotov's visit to London in May. The GAF was randomly reporting the towing of unidentified vessels off the south coast early in June, and on the afternoon of 11 June spotted a large concentration of shipping north-east of the Isle of Wight – almost certainly the force assembling for Exercise Yukon I. Von Rundstedt's immediate reaction to the series of photographs after 20 June was to drive home to the British that their intentions were patently obvious. He, too, was struck above all by the apparent neglect of any attempt at camouflage or concealment, suggesting the possibility of bluff or deception. On the other hand, there had been a recent increase in cable and railway sabotage; and on 22 June, twelve leaders of a Resistance organization had been arrested as they were in the act of preparing for a general sabotage operation, possibly in conjunction with a landing. Having no direct power of command over naval and air forces engaged in strategic warfare until a landing on the coast actually took place (the Navy in any case lost no time in making it clear on 27 June that there could be no question of offensive operations against the embarkation ports), the C-in-C requested 3 Air Fleet to concentrate their attacks on shipping.[2] Therefore, when the four FW 190s attacked units of the Rutter force off Yarmouth, Isle of Wight, early on 7 July, it was to make a point. Two LSIs loaded with troops had narrow escapes, HMS *Princess Josephine Charlotte* having a bomb pass through the messdeck and engine-room and explode under her keel without causing serious casualties but leaving the ship 'unable to steam', while *Princess Astrid* was hit on the starboard side by a bomb that fortunately exploded in the water. The fighter-bombers did not identify their targets as specifically connected with landing operations, claiming instead to have sunk a 10,000-ton steamer and damaged another of the same class and a motor minesweeper. A close watch was nevertheless kept up on English Channel ports and beaches, and the RAF logged 70 German aircraft between 7 and 13 July 'engaged in, suspected of, or capable of undertaking a recce flight' over the south and south-east – double the number for the previous week.[3] Remarkably enough, though, the next comprehensive PR survey was not flown until almost the end of July.

After reading von Rundstedt's situation report of 22 June, Hitler expressed the fear that his forces in the West were indeed too weak; as C-in-C of both the Armed Forces and the Army, he alone had the authority to move reserves from one theatre of operations to another. In Directive No 40 of 23 March he had laid down the guidelines for the defence of the European coastline against an enemy landing in force, and twice in April assigned a higher priority to measures to achieve this end than to urgent requirements in Russia and the Mediterranean. But in the spring Hitler was mainly preoccupied with Norway and Denmark and was busying himself as late as June with the anti-tank defences for a coastal battery at Hantsholm on the tip of Jutland.[4] The results of the aerial reconnaissance abruptly shifted his attention to the French coast, so that even before von Rundstedt submitted a second report on 27 June, Hitler had been galvanized into taking steps, some of which foreshadowed the provisions of his Order of 9 July. On 25 June he approved an OKH proposal to convert 23 ID, already on its way from Russia to France, into a Panzer division; while the three Panzer divisions already in France (6, 7 and 10) were to remain there, as were 7 Flieger division and the Hermann Goering Brigade, which was to receive an armoured component. The Navy was to keep a reserve of U-boats ready for intervention in case of a sudden assault.

On the following day, Hitler earmarked the re-activated SS motorized division Das Reich for France, opening up the prospect of setting up a Waffen-SS Panzer Corps should SS Leibstandarte Adolf Hitler and SS Totenkopf be moved to the West as well. On 29 June, the day after the start of the summer offensive in Russia, he presided over a small conference attended by Reichsminister Speer, General A. Jacob, the Army's leading expert on fortifications, and Major-General K. Zeitzler, Chief of Staff to C-in-C West, at which he began by expatiating on the threat of an Anglo-American Second Front forced on the enemy for political reasons. And on 12 August, he proposed to move a Corps of 6 Army to the West after the capture of Stalingrad; SS Adolf Hitler had already enjoyed a well publicized reception in France at the end of July, and the motorized division Grossdeutschland was earmarked as a further reinforcement in Directive No 45 of 23 July.[5] Thus the Führer's thinking about a possible Second Front in 1942 had been crystallized for the rest of the summer by the discovery of the really quite modest collection of LCTs, LCAs, R-craft, motor-boats, dumb barges and coasters between 20 and 23 June.

Why was Hitler so susceptible to the threat of invasion in the West? First of all, he was guilty of badly overestimating the possible scale of such a cross-Channel operation. At a meeting on 2 August, he maintained that the British could put ashore 300,000 troops in the two or three days it would take to collect German reserves, a total that could have been arrived at only by assuming that all 2,802 craft on the English coast were landing craft and by including warships as troop transports, as in the German invasion of Norway. Each destroyer, he projected, could land from 8,000 to 10,000 fully-equipped troops – a ridiculous figure even if extended over three days.[6] British amphibious capacity was in fact on a decidedly more modest level in 1942. Mountbatten told the COS in May that he could put ashore in France 4,300 troops and 160 tanks in a simultaneous lift; such an effort would require all the landing craft available – in other words, including those at training bases in Scotland – and it would take 21 days to land 132,000 troops, an estimate that dismayed Churchill at the time but that had not been radically improved by August. Quite apart from that, according to an authoritative estimate on 15 June, the maximum British force that would be available for operations on the Continent by 1 September was only six divisions, plus base and lines of communications troops. Ironically enough, it was Hitler who often accused his generals of overestimating the enemy. But in this case the end result was that the Army in France and the Low Countries was over-insured from July to the end of the year, even though Hitler stated on 2 August that it was still eight to ten divisions short of providing an unbroken line of defence.[7]

Secondly, Hitler and the High Command still had every reason to be confident that Germany was winning the war in the early summer of 1942, what with the fall of Tobruk and Sevastopol and expansive talk about pincers meeting somewhere near the Persian Gulf. Hitler may well have believed on 9 July that the only way he could now lose the war was by allowing the Anglo-Americans a foothold on the continent. Colonel-General Franz Halder, Chief of the Army General Staff, could deride the Naval Staff on 12 June for dreaming in continents, but even he was optimistic about German prospects in July. Goebbels, too, was optimistic but also careful not to repeat the mistake made in 1941 of claiming victories in advance. Hitler told Mussolini at the end of April that the land war would be decided in Russia and that a German victory there would rule out any chance of a Second Front and bring the war to a speedy conclusion.[8] At the same time, however, he soon came to appreciate that Germany could not operate decisively on more than one front at a time, and by 2 August he was talking in terms of an Atlantic Wall with coastal batteries heavily protected by concrete and strongholds capable of holding out for weeks on end. At another meeting on the same subject on 13 August, Hitler projected a wall of 15,000 bunkers sheltering 300,000 men, to be completed by May 1943; by then, with the Soviet Union cut off from its raw materials, Germany would be in an unassailable position. He had known exactly how strong the West wall had been in 1939 but he was now having difficulty in estimating the strength of 'field defences' in the West and therefore in calculating how many reserves it would be safe to pull out at a critical moment. At yet another session on 29 September, after the Dieppe raid, Hitler held forth for three hours on the urgency of building up an Atlantic Wall before the Allies had a chance even to attempt a Second Front.[9]

Thirdly, there can be little doubt that Hitler considered that his political prestige would be at stake in the event of a major landing in the West; significantly enough, he announced in his Order of 9 July that he would personally take command of the anti-invasion battle. Hitler paid careful attention to the daily digest of the foreign press and wire services prepared by Otto Dietrich, which must have reinforced his sense of the propaganda value of raiding and the importance of German counter-propaganda. Indeed he took the raid on St Nazaire almost as a personal affront; there was even mention, briefly, of a retaliatory raid on the English coast. After Dieppe, Hitler played up the idea that it had been a poorly prepared operation with ambitious objectives – to capture Le Havre or take the Channel defences from the rear – that had been forced on Churchill by Stalin in Moscow less than a week earlier; at the meeting on 29 September, he chided those who had earlier thought him a 'seer of ghosts'. One of the principal reasons for expecting further raids after Jubilee was the loss of British prestige; von Rundstedt predicted another raid as early as 21 August, *aus Prestigegründen*.[10]

Actually, although Hitler took pride in always distinguishing between 'hard fact' and fantasy in propaganda, including German propaganda, there was a close correspondence between what is known about his innermost thoughts about a Second Front and the official line followed by Goebbels. In the first half of 1942, German propaganda brought up the subject of a Second Front only for the purpose of ridicule, with Hitler answering Beaverbrook's bullish speech on the subject in New York on 23 April with a taunting invitation to the Allies to 'have a go'; Goebbels was equally scornful of 'grandiloquent propaganda' in his articles in *Das Reich*. But on 5 July Goebbels abruptly changed his tune: now the possibility was opened up that the Allies might make an ill-advised attempt – 'in a fit of military madness', no less – to bail the Russians out. The German public was led to expect an invasion in the West by an enemy torn between a desperate military gamble and an internal political crisis stirred up by Communists and fellow-travellers. At this stage, Goebbels was all but 'committing' the Allies to an immediate Second Front, but by the end of July there was a subtle shift in emphasis to the strengthening of German forces in the West; widespread publicity was given to the parade of SS Adolf Hitler through Paris on 29 July and to the volume of concrete being poured into the Atlantic Wall. The exaggeration of German strength was in accordance with what had been done earlier in the year in Norway and with what little strategic deception Army Group D attempted in France. The British were allegedly taken in 'for a time' by having 6 Pz in Brittany display the flag of a Panzer Corps and by assigning the division an area covering the whole peninsula. Goebbels told his staff on 28 July that Germany had nothing to gain by provoking an invasion in the West; rather, the British should be given the chance to formulate a convincing excuse to explain to Stalin why they were failing to honour the IOUs he had been given in May.[11] Far from this being a case of the Führer's swallowing his own propaganda, he and the various organs of German psychological warfare more or less simultaneously decided to take the threat of a Second Front seriously – eloquent testimony to the impact of the PR missions flown between 20 and 23 June.

Few, if any, officers of sufficient seniority to form an independent strategic judgment agreed with Hitler's reading of the situation in the West. As Warlimont saw it from his vantage point at OKW (Operations) – Hitler's immediate staff – the Führer had lately succumbed to considerations of *Sicherheit überall* – safety at every turn – and had lost whatever sense he had possessed of concentration at the decisive point – *Schwerpunktbildung*. Halder, unlike Hitler, was still prepared to take calculated risks, a virtue Warlimont attributed to his General Staff training. Warlimont may have had a point, given Hitler's earlier readiness to use the West wall essentially as a bluff during the campaign in Poland.[12] The Army General Staff at Oberkommando des Heeres (OKH) were mainly preoccupied with the war in Russia, and Halder left any reference to Dieppe out of his diary; the Führerhauptquartier, he complained, was dominated by instant reactions to momentary developments, among which he would have included the transfer of forces to the West. He and Jodl, for example, joined forces in opposing the transfer of the Grossdeutschland division to France, but Hitler gave in only when the pressure of events on Army Group Centre's front became irresistible. Halder and Jodl took their stand on the imperatives of the Eastern front, which should have been a more effective argument than the hollowness of the threat in the West.[13] During the first serious crisis of the summer offensive, Hitler had been anxious to use mobile formations to pursue and entrap Russian formations retreating south of the Don, whereas Field Marshal von Bock, C-in-C Army Group South, was intent on the capture of Voronezh just to the east of the river, using a Panzer division and Grossdeutschland to do so. Their tense arguments culminated in von Bock's dismissal on 15 July. Significantly enough, even while in the grip of these constraints, Hitler was reluctant to release SS Adolf Hitler to 1 Pz Army on 5 July for an operation in which, as one historian has put it, 'its mobility and motivation would have been a considerable asset', because he had already decided that it should go instead to France. The tragedy from the German point of view was that the Führerprinzip frequently reduced top-level strategic discussion to haggling over the transfer of single divisions.[14]

At HQ Army Group D, von Rundstedt was never very suggestible on the subject of a Second Front. He took pride after the war in not having been deceived by Starkey in 1943, and his initial reaction on 22 June 1942 was to smell a rat.[15]

Nonetheless, given the overall weakness of his forces and the lack of anything but the sketchiest of defensive preparations when he took command in March, he was in no position to decry the danger of large-scale operations, no matter what his instincts as a soldier told him. He began by proclaiming the prevention of a Second Front as the historic duty of the troops in the West and appealing to them for vigilance and improvisation until reinforcements arrived. But in a comprehensive report on 6 June he was still obliged to describe his Army Group as inadequately trained and badly equipped. The strategic reserve of armoured and motorized formations on which he relied to complete the break-up of a large-scale landing impeded by fixed defences on the coast still existed only on paper; coastal divisions had to cope with green recruits and captured equipment; artillery was weak and naval coastal batteries were often badly sited from a defensive point of view; many officers were too old to face the strain of continuous fighting over a prolonged period. The OKW Operations staff conceded that C-in-C West's appreciation was accurate enough but considered his forces adequate to meet the limited operations that were all the Allies were in a position to attempt. Halder, on the other hand, thought Army Group D equal to any situation likely to arise in the West – this too in June, before the reinforcements arrived. In any case, von Rundstedt continued to deploy his forces in an anti-raid rather than anti-invasion configuration. By the start of August he was reportedly convinced that, even if the British gained a foothold, they would very quickly be thrown out again; but he reached the stage of formally conceding the adequacy of his coastal defences only on 1 September. Raids, in his opinion, were to be expected at any time because they were cheap and promised some propaganda return, reasons for which they were thought to appeal to the British mentality.[16]

Von Rundstedt's Chief of Staff later lent his name to the propaganda fiction that Jubilee had been an abortive Second Front, but he did so largely for careerist reasons. Zeitzler told a visiting staff officer on 26 or 27 June that he expected a landing in July because the purpose behind such a large collection of boats could hardly be mere bluff. The likeliest targets were the U-boat bases in western France. Such an operation, as he described it, was neither a *Grosslandung* nor a raid but fell into the intermediate category known to the Germans as operations with limited strategic goals – *mit begrenztem Operationsziele*. At this *operativ* or operational level, the enemy would attempt to inflict appreciable damage on the German war effort by capturing a strongpoint or disrupting air or naval operations for some length of time, as in the putative attempt to seize the Brittany peninsula outlined by Hitler on 23 March. Although such operations would require strong forces and were fated in some cases to end up with a difficult re-embarkation, they were taken seriously by all levels of German command in 1942. The British had no corresponding operational category between the strategic and tactical, dividing operations rather more pragmatically between those where they intended to stay (invasions) and those where they did not (raids). Zeitzler was quite confident in June, and his confidence was evidently unshaken by attendance at Hitler's small conference on 29 June. He assured Grand-Admiral Erich Raeder, C-in-C Navy, on 3 August not only that there would be no *Grosslandung* but also that the U-boat bases were all but impregnable, each one being defended by an infantry division with mobile reserves positioned to intervene on the day of a landing.[17]

The question of a possible Second Front should of course have been one for Intelligence to deal with. But Intelligence in the higher echelons of the Third Reich, as Martin Geyer has pointed out, was used almost as much as a weapon to establish credibility in Hitler's eyes as it was to assess enemy intentions. No-one realized this better than von Ribbentrop, the Foreign Minister, who told the Japanese ambassador in April and again in June that the Allies were preparing a Second Front. Not surprisingly, the Foreign Minister reacted promptly to the discovery of the preparations on the English south coast and instructed all embassies and consulates on 29 June that their most important task was to gather and pass on at once all reports about when and where the Anglo-Americans might launch an invasion. Clearly he wanted to be first with information about the most pressing question of the day, but the flood of reports he received in reply to his circular, largely consisting of the predictable collection of half-truths and rumours, was *not* to be shared with the armed services. There could be no better example of the fragmented and competitive state of 'Intelligence at the Top', where there was no Ic agency worthy of the name at the OKW level and where Hitler was never at a loss to provide himself with selective 'confirmation' of his strategic hunches.[18]

Although Halder and Jodl were almost as intuitive in rejecting the possibility of an invasion in the West as Hitler was in sponsoring it, this is not to say that German Military Intelligence failed to take a stand on a strategic question of such importance (and novelty). Unfortunately, many of the basic documentary series for 1942 are missing, but at least some significant individual documents have survived. Long before the June reçonnaissance, Liss and his staff at Foreign Armies West had reached agreement with Naval Intelligence that there was nothing to fear from a *Grosslandung* in 1942. Liss submitted a comprehensive strategic forecast on 1 March ('Die britisch-amerikanischen Operationsmöglichkeiten gegen Europa und Afrika im Jahre 1942', Generalstab des Heeres, Abt Fremde Heere West, Nr 632/42, gKdos Chefs), which boldly argued that the Allies would restrict themselves to a number of separate operations in 1942 so as to conserve manpower for a maximum effort with newly-built tonnage in 1943. The shortage of ocean-going tonnage would force the postponement of a major cross-Channel initiative, although there were enough air and military forces in England, over and above those required for defence, to launch it. The steady reinforcement of the Middle East via the Cape that started at the end of 1941 at the rate of roughly two divisions a month was a maximum effort, as indicated by a further reduction of civilian rations in January; ships leaving in mid-December could not be back in home ports until the beginning of May. Foreign Armies West was following this process carefully, using Sigint to mark the disappearance of units from wireless nets in England and reports from Abwehr agents about convoy movements. Liss clearly found cause for self-congratulation in that his appreciation was the result of some rare collaboration with Naval Intelligence (3 Seekriegsleitung) and even his 'old adversaries' in the Abwehr.[19]

It was once again characteristic of the German version of 'Intelligence at the Top' that Liss's appreciation of 1 March earned him an instant rebuke from Hitler. What offended him was not Liss's data but his conclusions, for Hitler had only just proclaimed Norway to be the 'fateful zone' of the war and the Channel Dash in February, followed by the dispatch of twenty U-boats to Norway in March, were only two indications of the importance he attached to strengthening defences in the North. Now Liss was threatening to undermine all such precautions by spreading a false sense of security. Henceforth care should be taken that such long-range strategic forecasts should conform to the Führer's thinking, above all for theatres of operation not subordinate to the Army General Staff (OKH) where they might contradict directives issued by OKW. The Führer's thinking was guided by insights of a political and psychological nature beyond the grasp of Intelligence colonels and their staffs, which was why the appreciation was ordered to be withdrawn from circulation. Foreign Armies West, incidentally, remained illogically subordinate to OKH in Berlin, although the West, including the Mediterranean, was an OKW theatre of operations; and Liss was restricted to visits once or twice a month to OKW in East Prussia, when, according to Warlimont, he was never allowed to see Hitler himself.[20]

Liss at least had the consolation that subsequent developments apparently substantiated his evaluation. Japan's entry into the war forced a redeployment of troops from the Middle East to India and Australia, and Rommel's offensive in May/June could not have been better timed from the tonnage point of view, leading to a new round of reinforcements for 8 Army. None of the shipping leaving British ports after mid-April could be counted on for a cross-Channel invasion later that year. Throughout the spring and summer, Liss did his best to point out that the British Empire could not survive defeat in the Middle East and lobbied for reinforcements to be sent to Rommel. On the day that Hitler issued his Order about the danger in the West, Liss forwarded to OKH/OKW an appreciation of the consequences for the enemy of the loss of Suez, in which he concluded that the British would be hard put to maintain logistical support for the forces they already had in the theatre. Liss later inspected the shambles on the beach at Dieppe, which he saw as confirmation of his forecasts: surely now there would be an end to nervousness about that invention of Allied propaganda, a Second Front.[21]

Three observations might be made about Liss's analysis. First of all, the Abwehr had not, as claimed, uncovered the timing and tempo of the WS convoys. A comparison of actual sailing dates with Abwehr reports recorded in the Naval Staff (Operations)

War Diary shows wide discrepancies. Non-existent convoys were duly reported at Freetown, Cape Town and Durban as they made their notional way around Africa. Moreover, Foreign Armies West underestimated the number of troops transported, an error which probably served Bevan's purposes, since he initiated steps in July to convince the Germans that the numbers of American troops arriving in the British Isles and of British troops departing the country were both greater than was the case. Sooner or later, of course, German Field Intelligence was bound to identify units recently arrived in the Western Desert.[22]

Secondly, it was no secret that the Allies were suffering from a shortage of tonnage. Apart from anything else, the U-boats were enjoying their most sustained success of the war in the first half of 1942. Liss decided in March that lack of shipping would restrict operations against the French Atlantic coast, the Iberian peninsula or north-west Africa, a confident and far-reaching conclusion that was unavoidably based on very sketchy data. Calculation of the merchant tonnage required for the initial assault, the build-up and the maintenance of Torch, including effects on the import programme for the UK, fully taxed the British Shipping Committee in the summer of 1942. Not surprisingly, therefore, the Germans got it only partially right and were taken badly by surprise at the scale of Torch.[23]

Thirdly, it would have been extremely difficult for British deception to dislodge the *idée fixe* of the shortage of tonnage. Gathering up 200 coasters for a Sledgehammer in September would not have altered Liss's opinion in the least. Hitler already believed in the possibility of a Second Front, so there was nothing to be gained by impressing him further, and even his best efforts were unable to prise the Grossdeutschland division loose from the Eastern Front. Therefore there was very little chance of increasing the diversion of ground forces to the West; indeed, according to opinion at OKH at the end of July, the British had already achieved the objectives of a Second Front by forcing the transfer of one infantry and four Panzer divisions to the West. Although Stalin would never have agreed – he thought more in terms of 40 divisions – this was still five more than the CIGS thought possible in March.[24] As for the GAF, it was always a difficult target for deception because it was not large enough to operate in strength in more than one theatre of operations at a time; lack of reserves by 1942 was making mobility a necessity rather than the asset it had been earlier in the war, but it was also good reason for the GAF High Command (Luftwaffenführungsstab) to pause before making strategic commitments. Thus even the potential effectiveness of cover and deception was quite circumscribed in the summer of 1942, partly because of the lack of a central reserve of German ground and air forces, and partly because there was no widely held German presupposition about a Second Front to build on.

It was one thing to take a confident stand on the question of a Second Front, quite another to make sense of the activities on the English south coast. Here the lead was taken by the Naval Staff (Operations) in a lengthy study of 20 July ('Feindlandung im Westraum', 1 Skl Ib gKdos Chefs 1363/42) in response to Hitler's interpretation of the strategic dilemma being imposed on the British by Germany's 'schnellen und grossen' victories in Russia. According to the Naval Staff, the Western Allies were pinning their hopes on the continued survival of the Soviet Union. Pressure was increasing to launch an emergency operation in the West, yet an even greater incentive for such an operation was the catastrophic course of events for the Allies in the Atlantic over the last six months. The monthly sinking rate of 800,000 BRT required to bring the British to their knees had already been attained. The Allies must take drastic steps soon to defeat the U-boats before they could so much as contemplate an invasion of western Europe in 1943. Bombing the bases had failed, so that the only recourse left was to neutralize them by airborne or amphibious assault. The Naval Staff hammered home to OKW what had become a familiar refrain, namely, the war-winning – *kriegsentscheidende* – potential of the U-boat campaign in what had become a war to control sea communications: all to no great effect, apparently, since Warlimont was struck by the lack of mention of the U-boat war in Greiner's notes for the OKW War Diary for the second half of 1942.

The British were seemingly doing their best to attract attention to their embarkation ports opposite the Bay of the Seine. Why? A *Grosslandung* was discounted as probably neither part of the enemy's planning nor within his current resources in tonnage. The concentration of landing craft and coasters in the central and western Channel was simply not powerful enough to launch an invasion; nor was it a direct threat to Brest, Lorient, St Nazaire or any of the other U-boat bases on the Atlantic coast. The only way to paint a plausible picture was to hypothesize some sort of 'main force' concealed somewhere in the Highlands of Scotland (for which there was some support in an Abwehr report originating in Stockholm on 9 July), plus a strong Anglo-American airborne force. Accordingly, it was decided that the chief purpose behind the all-too-evident activity on the south coast was to distract German attention from Brittany. By threatening and then launching an attack on the coast of Normandy, the Allies were hoping to draw off German reserves from the west and at the same time shield the approach of the main force. Once parachute and air landing troops or Commandos had taken one of the U-boat bases, the main force convoy would disembark troops in strength and the other U-boat bases would then be reduced from the landward side. This *Hauptlandung* was envisioned as an operation to destroy the bases but also to gain a foothold on the Continent, albeit with limited objectives at first – 'operativ', as opposed to a *Grosslandung* with far-reaching strategic goals. Alternatively, of course, the forces in the south of England might be the nucleus of a larger army charged with the mission of seizing a bridgehead in Normandy from which mobile units would sally forth against the U-boat strongholds. This super-raid was a variation on earlier speculation about landings on the two coasts of Brittany to cut the peninsula off altogether. But there was the obvious drawback to a drive from the north-east of a long and vulnerable flank. On 22 July, at any rate, the Naval Staff reiterated the urgency of extending GAF reconnaissance to cover the Bristol Channel, Irish Sea and North Channel. Plans were laid for a U-boat ambush of the main force on its way south, and Raeder agreed to carry out a personal inspection of naval defences in the West so that Hitler could be reassured that everything humanly possible had been done to strengthen them.[25]

What the Naval Staff had done, so to speak, was to invent a cover plan to explain what was going on across the Channel. In a sense they were driven to self-deception because they were getting little or nothing to work with from their agents in England. The British were well aware of German sensitivity to any threat to their U-boat bases and yards, which was exactly the kind of presupposition upon which successful deception depended. Martian Report No 2 stated on 9 June:

> In general it may confidently be stated that the only vital objective in the West at the moment are naval bases, particularly submarine bases. Short of losing the whole of the occupied territories, only the loss of naval bases is serious enough, in the German view, to affect the course of the war.[26]

There were actually the makings of a promising cover plan for a major cross-Channel operation in 1942 in the shape of a real Expeditionary Force that had been forming in Scotland since February as a permanent organization to undertake overseas operations. Even so, the direct benefits of such a cover plan for a raid like Jubilee would have been doubtful. The German Naval Staff recommended that no troops should be withdrawn from Brittany in the event of a landing on the Channel coast; that would have been the extent of the cover plan's contribution to the success of, for instance, an operation to seize the Cotentin peninsula or Le Havre. But a delay in the movement of strategic reserves could hardly have contributed much to the success of a raid lasting a matter of hours. The German Army's deployment of reserves within easy reach of Brittany, according to the Martian Report issued on 19 August, showed that they basically accepted the Navy's analysis:

> The presence of this concentration [6 Panzer, 17 ID, SS Das Reich, General Goering Brigade] suggests that the enemy is more nervous of an Allied attempt to seize and hold the Brest peninsula than of other possible Allied operations.[27]

German naval staffs first gave serious consideration to cross-Channel operations in the spring and summer of 1942. At the request of the GAF, the Operations division of the Naval Staff compiled a comprehensive memorandum (Seekriegsleitung, B Nr 1 Skl I op, 16934/42 gKdos) which was distributed on 11 July, complete with an analysis of geographical and operational factors and covering all categories of assault from minor raid to invasion. The best place by far for an invasion was found to be the Bay of the Seine, which offered the most sheltered beaches of any stretch of the French coast, not to mention proximity to Le Havre or Cherbourg whose capture was considered essential to handle large merchant ships and ensure the flow of supplies whatever the weather; a few months later, the British invasion planners in 1942 reached exactly the same conclusion.[28] Naval Group West and Vice-Admiral Ruge, Commander Naval Defences West, had reached it in May. Le Havre and Cherbourg both had airfields nearby and could not quickly be made unusable. The English coast opposite the Bay of the Seine had a large number of small estuaries and harbours not easily covered by aerial reconnaissance; even in mid-summer a vessel making 12 knots could make the crossing under cover of darkness; and once a toehold had been secured, a shuttle system of supply could be organised under air cover with which the German Navy would be powerless to interfere. The coast to the east had little to offer because Fécamp and Dieppe were of limited capacity and could easily be blocked. Further up-Channel, according to Ruge, the enemy would benefit from the shortest crossing under maximum fighter cover but ports like Boulogne, Calais and Dunkirk suffered from the same disadvantages as Fécamp and Dieppe, plus navigational difficulties. An invasion of the Pas de Calais would be forced to use embarkation ports in the Thames estuary and so run the gauntlet of the heavy batteries between Calais and Boulogne. As Captain Bramesfeld, commander of 2 Defence Division, pointed out, there could be no prospect of landing heavy equipment on the Belgian coast; any landing to the east of Le Havre would be for diversionary purposes only or to neutralize a specific strongpoint. The results of the PR on 20–25 June merely solidified this appreciation on the part of naval authorities and explains why Ruge gave top priority on 28 June to increasing patrol boat (*Vorpostenboote*) stations in the Bay of the Seine.[29]

Attempts were made, naturally, to fine-tune these fairly basic appreciations. The pattern of the enemy's offensive minelaying, for example, might offer clues to his future intentions, so monthly surveys were begun by Naval Group West in May. The British were found to be concentrating on the Straits of Dover and the approaches to the U-boat bases to the west, particularly in the case of ground mines which were more difficult to sweep quickly than moored mines. Brest was rather a puzzling exception. No ground mines and few others had been laid in the Bay of the Seine, a pattern that remained constant for June and July. But Naval Group West shrank from drawing the obvious conclusion that the Bay of the Seine seemed to be a prohibitively good bet as the choice location for a major landing. Surely the enemy would not draw attention in this way to the very area he intended to attack? With their well-known fondness for deception, the British would be more likely to mine the target area with mines fitted with a timing device so as to gain surprise by suddenly approaching the coast over a field of disarmed mines. The Germans themselves had restricted naval minelaying off the sectors of the English coast selected for Sea Lion to ZE (*Zeiteinrichtung*) mines until April 1942. Somewhat absurdly, given the weight of evidence to the contrary, anxiety arose about a possible landing in the southern part of the Bay of Biscay when a single mine with a timer was salvaged early in June.[30]

Attempts to define the all-important question of timing had little more success. On 10 July, Hitler asked the naval representative at OKW for the likeliest dates in July and August for a landing on the Channel coast. The Operations division of the Naval Staff replied that the British would probably land on a rising tide about two hours before daybreak and during the half-moon phase, conditions that prevailed in the Bay of the Seine from 5–7 August and 4–5 September; the Japanese staged their assaults along these lines. In fact, the British themselves considered that the ideal timing of an assault called for an infantry landing an hour and a half before first light on a morning when the tide was such that LCTs arriving at first light would beach near High Water and so be able to refloat at once. On 13 July, Captain Wagner, chief of the Operations staff, ordered the hydrographic and meteorological branch to prepare diagrams showing conditions – twilight, phases of the moon, tides and currents – that might determine the most probable dates for landings. Hitler then requested the Naval Staff to calculate from those charts the likeliest upcoming dates, based on the requirements observed in the planning of Sea Lion. But the outcome of this inquiry was considered too academic to be distributed to naval commands, given the very different conditions the British were expected to consider necessary.

Not surprisingly, therefore, military authorities miscalculated the periods of *drohende Gefahr*. Jubilee, for example, was planned so that it could be carried out on 18 August or, in the event of postponement forced by the weather, on any subsequent date up to and including 23 August, whereas the German period of alert began on 10 August and expired on 19 August. By the same token, the period of suitable tidal and astronomical conditions for Rutter ended on 9 July, a day before the start of what the Germans took to be a period of increased danger. To complicate matters further from the German point of view, the Jubilee assault was scheduled to take place on a falling tide on 18, 19, 20 and 21 August, with re-embarkation on the next rising tide – on 19 August, High Water was at 0403 followed by Low Water at 1056 – but on 22 and 23 August the tanks were to be landed and re-embarked on the same rising tide, with the infantry following after High Water. Some German interrogators of Canadian POWs interpreted 'two-tide' on 19 August to mean that a second wave of British troops would have landed had the initial assault been a success.[31]

To the extent that they paid any attention at all to the question of an invasion in the West, the OKH tended from the start to favour the Pas de Calais as the likeliest place for a landing. After the raid on St Nazaire, Halder briefly showed some interest in the possibility of large-scale operations against the coast of Occupied Europe. He was not satisfied with an assurance from the Naval Staff on 3 April that a landing in Holland or France was impossible because the British could spare enough tonnage for only two divisions. It was well known, he pointed out to the naval liaison officer to OKH, that specialized landing craft were being built in the UK and US, so that the enemy might already be in a position to land 10 to 15 divisions between the Scheldt and Boulogne or the Seine and Brittany. He was informed in reply that there were not enough landing craft to land a superior force in Holland or France; although it was left vague just how many craft of the three types that had been identified (LCA, LCM and LCS) there were in the UK, Halder was evidently satisfied with this assessment.[32]

Exactly how much German naval and military Intelligence knew about British amphibious techniques and capabilities before Jubilee is not easy to say. There were no models from PR to measure the scale of assault represented by greater or lesser concentrations of craft and shipping. Such raids as had taken place were not on a sufficient scale to provide visual or other signs of preliminary movements. Their own plans for Sea Lion were of limited use to the Germans because the British had a far greater variety and number of vessels at their disposal than the 1,800 barges and assorted tugs, motorboats and coastal steamers assembled by the Wehrmacht. Little headway had apparently been made since then, judging by the critique provided by the Japanese Military Attaché of landing exercises held at Antwerp in the first half of July 1942, a critique that covered everything from lack of a proper doctrine to lack of rubber shoes – *gegen Ausgleiten beim Anlandspringen*. In April the Naval Staff confessed that their intelligence about 'Spezial-Landungsboote' was very patchy; they knew, for example, that the LCA could be carried in davits and had salvaged a 'Sturmboot' at Maaloy (which exploded in Bergen harbour in May). The Abwehr put the total of available landing craft at 200/210 and reported in mid-April the appearance of 120/140 of them at Poole and 30/36 at Rye, for the first time on the south coast. Captured documents brought the first solid information about the specifications for the LCT later that month. On 18 May, 3 Seekriegsleitung circulated a report on the Motor Landing Craft (LCM), and at the end of the month the translation of a British manual on landing tactics reached the coastal IDs, with a description of the LCA, LCS, LCM and the new LCT. Fairly frequent references appear in military documents, including Hitler's Directive No 40, to armoured landing craft and the need for anti-tank guns and ammunition to counter them. On 6 July, 302 ID received copies of photographs from the *Illustrated London News* of 2 May of R-craft on manoeuvres in the US and British landing and coastal craft.[33] Generally speaking, though, von Rundstedt was being frank when he admitted in his Battle Report that the enemy had significantly more modern landing equipment than had hitherto been supposed.

Until the documents and equipment captured at Dieppe presented them with a priceless windfall, the Germans could not be sure how all the parts of a major assault fitted together. 'Sturmboote' might come across the Channel in an enormous swarm or they might be towed or carried in a 'Mutterschiff'.

Mountbatten took particular trouble to conceal from the Germans the fact that the LSI *Prince Albert* had taken part in Bristle and to prevent any photographs of LSIs or references to them from appearing in the press before Rutter. The only Mutterschiff to figure prominently in German documents was the *Dewdale*, technically speaking one of three Landing Ships (Gantry) run by the Ministry of War Transport and equipped to carry landing craft (usually LCM) to be used for landing stores after an assault. The Abwehr made the most of an opportunity to observe the *Dewdale* at Gibraltar in February. At 17,000 tons, the *Dewdale* was not a 'mother ship' in the same sense as the smaller, faster LSIs operating in the Channel, whose role in a landing COHQ was so anxious to keep from the enemy. The Germans were certainly aware of the latter, since Foreign Armies West reported the presence of two 'Fahrschiffe' converted for landing purposes at Falmouth and four more at Newport on 15 August.[34] The naval analysis of the captured documents ('ENGLAND: Das Zusammenwirken des 3 britischen Wehrmachtteile in der Planung und Durchführung des Landungsunternehmens Dieppe am 19. August 1942', Anlage zu 3 Abt Seekriegsleitung, B Nr FL 18440/42 g, 8.10.42) not only cleared up any remaining obscurity about the role of the LSIs ('Landungsboote-Transportschiffe') but also corrected the badly distorted picture of the operation that German staffs had put together in its immediate aftermath. Interestingly enough, 3 Skl commented that in a *Grosslandung* the British would be sure to use all available LSIs; the nine used at Dieppe represented only 50 per cent of the ships in their class that had been known to be operational before 19 August, which suggests that previous intelligence about the converted Dutch, Belgian and British cross-Channel steamers and the 'Glen' ships might have been better than the surviving documents allow. It was known, too, that the *Dewdale* was one of three ships in her class. A puzzling reference to the *Ulster Monarch* and *Royal Scotsman* as 'Port Invasion Ships' to force harbour entrances – they were in fact LSIs – can be explained, as so often in German sources, by its origin in an *Agentenmeldung*.

In short, before Jubilee the Germans had at best a general idea of likely British tactics for a major landing. Von Rundstedt's Basic Order No 13 of 21 July, which tried to describe those tactics, was limited to pointing out that parachute troops would be assigned to capturing bridges and airfields, saboteurs would be active, aircraft would launch low-level attacks on marching columns of reserves, as they had on recent exercises in England, good use would be made of smoke, and so on. After Jubilee, the Germans found themselves in possession of two-thirds of the Detailed Military Plan (JG One), the missing third being largely accounted for by Appendix C, 'Allotment of Personnel, Equipment and Stores'. Why it was thought necessary to take ashore parts of the orders dealing with the mounting of the operation up to the point of embarkation is impossible to understand. Of no less value to the enemy were significant portions of the Naval Operation Orders (NFJ 0221/92), of 10 August, complete with maps and schematic drawings. JNO 1, with appendices, gave an overview of the operation, while other captured orders covered the withdrawal (JNO 6), the role of coastal forces (JNO 7), the use of smoke (JNO 8), the passage through the minefields (JNM/S), and the movements of landing craft (JNLC). The only serious omissions were JNO 2 (orders for LSIs) and JNO 3 (orders for destroyers).[35] Considering the number of German reports dealing with Dieppe that found their way into Allied hands before D-Day, Jubilee was outstandingly well documented in the slightly unusual sense of the exchange of captured paper.

During the exchange with Halder in April, the naval liaison officer to OKH told him that the only way to keep track of the movements of landing craft and other vessels was by aerial reconnaissance. There should be a routine survey once a week of the English south and south-east coasts; at the first sign of a concentration of shipping and small craft, the GAF should fly daily PR missions to cover the harbours from Falmouth to the Wash. Long-range reconnaissance was already the source of much inter-service friction. After the raid on St Nazaire, for instance, Raeder and his staff found themselves called to account by OKW for the surprise achieved by the enemy. They in turn felt badly let down by the GAF's failure to spot the Chariot force on passage or to join in the harassing of its remnants as they made their escape the following day. Then there was the surprise appearance of the Myrmidon force off St Jean de Luz on 5 April, leading the Navy to petition for regular evening reconnaissance up to the British coast in the southern part of the North Sea and Channel and up to 150 nautical miles offshore along the Atlantic coast.[36]

CHANNEL WATCH 1942

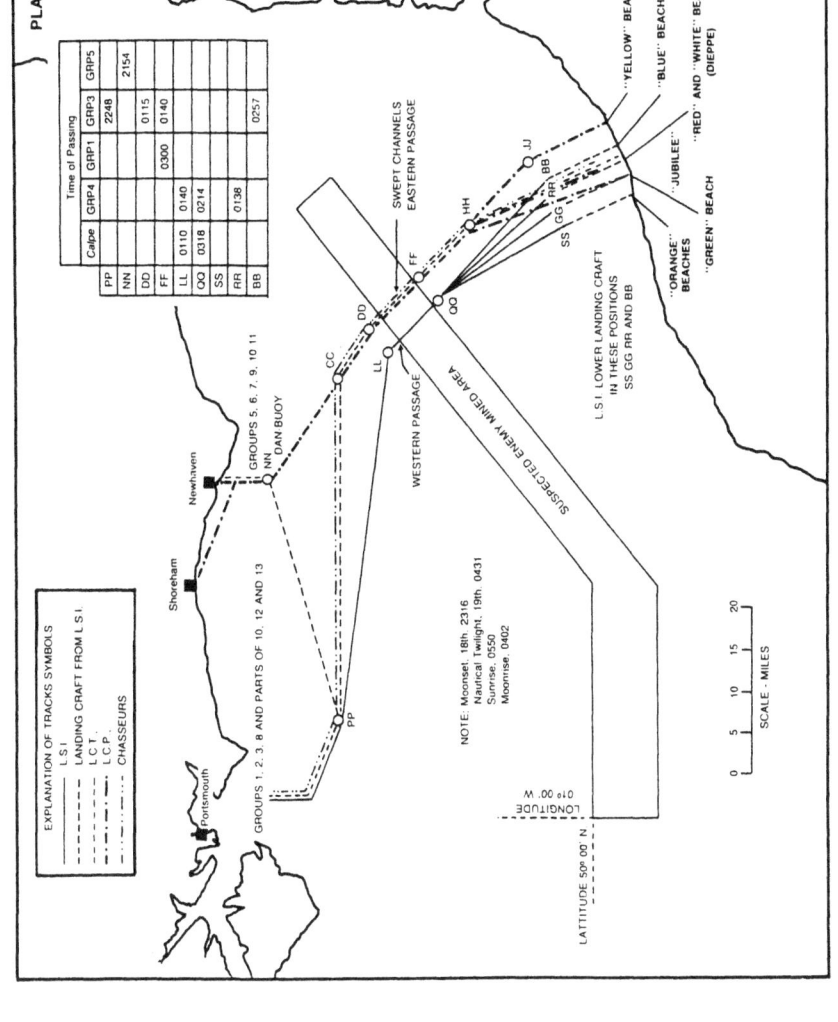

Map 1 Operation Jubilee (Crown copyright)

Surveillance on this scale was in fact far beyond 3 Air Fleet. Visual reconnaissance of harbours and beaches on the English south coast increased markedly in the first half of July. But the next PR survey did not take place until 28–31 July, giving a rate of once-a-month instead of the Navy's requested once-a-day. No appreciable changes were discovered in the strength and disposition of the pocket armada, which was numbered at 2,435 ships and other craft and 725 landing craft. How the Germans arrived at such a total for landing craft is hard to say, because there were only just over 400 of all types in Home Waters on 13 June and only about 165 were used for Jubilee. Most likely, some of the 900 dumb barges (possibly those fitted with bow ramps), were being counted as landing craft. The Air Staff was aware of the Navy's requests for coverage of the Bristol Channel, Irish Sea and North Channel, but had done nothing about it. None of the surviving documents mentioned the importance of reconnoitring airfields as well as anchorages, although Hitler had alluded on 25 June to the danger of an airborne assault. On 1 July, the Naval Staff War Diary mentioned that a 'planned assault' on RAF bases had been cancelled after the transfer of two bomber groups from 3 Air Fleet to Sicily.[37]

The next systematic PR sorties were not flown until after Jubilee, 24–28 August, when it was admitted that no comparison was possible between totals for the end of August and those for the end of July for the coast between Portland and Beachy Head. Almost unbelievably, the most crucial sector (rather misleadingly described as 'small parts of the Southampton area') had not been properly covered since 23 June! Much of the value of PR, of course, rests on comparing photographs taken at frequent intervals. Thanks to better photography over a wider area, the August missions produced a total of about 5,000 small ships suitable for landing operations – 3,296 on the south coast and along the south shore of the Thames estuary, plus an estimated 1,543 in the western part of the Channel. The estimate was taken from the July return, because this time it had not been possible to cover the coast west of the river Axe. Again, nothing was said about the Bristol Channel and Irish Sea, although coverage had been promised in July once the high-altitude Junkers (Ju) 86P arrived. This was a bomber-recce aircraft with a pressurized cockpit, one of which was plotted for the first time over the south of England on 24 August. Sightings of a high-flying aircraft were reported on 25 and 28 August, and two Spitfire IXs spotted an aircraft at 41,500 feet over the Isle of Wight on 29 August. No RAF fighter could make an interception at such an altitude. On 12 September, a specially modified Spitfire IX intercepted but did not shoot down a Ju 86P over Salisbury.[38]

The reasons for 3 Air Fleet's feeble PR effort can only be surmised. It was partly a reflection of the general decline of the GAF in the West as an offensive force. A few long-range bomber groups, mostly based in Holland, had been engaged since April in the retaliatory bombing of British cities other than London – the so-called Baedeker raids. The most formidable force in 3 Air Fleet's order of battle was defensive – the two fighter Geschwader (JG 2 and JG 26). RAF Fighter Command admired the skill shown by two Ju 88 bomber-recce units of IX Fliegerkorps, who were so familiar with the limitations of the reporting system between the Thames and the Wash that it was estimated that only a quarter of their activities were plotted. But they were anti-shipping specialists. Overland daylight reconnaissance at 20,000 or 25,000 feet must have been a dangerous proposition in face of the efficient warning system on the south coast and the enormous numerical superiority of the RAF, if not yet as deadly as it became in 1943 and 1944. Lack of fuel, bad weather in July and unwillingness to risk such a scarce and valuable aircraft as the Ju 86P all probably contributed. It is also quite possible that the GAF High Command took the possibility of invasion less than seriously, although lack of documentation makes it risky to generalize. Certainly the Chief-of-Staff of 3 Air Fleet told Raeder on 4 August that the situation on the English south coast was being carefully watched; in his opinion, though, the British were bluffing or possibly parking newly constructed landing craft between Plymouth and Portsmouth for want of anything better to do with them. Martian Report No 14 of 2 September is perhaps worth quoting on this subject:

> The German Air Ministry, whatever conclusions it may have drawn from the air battle over Dieppe, remains, according to reliable sources, quite as scornful as ever of the possibility of an Allied invasion in the West.[39]

To talk in terms of a concentration of shipping, barges and landing craft extending from the Thames estuary to Falmouth was clearly nonsense. It was essential for German Intelligence to determine from the special maps locating the various types of craft where the true 'Schwerpunkt' lay. As early as 25 June, von Rundstedt was zeroing in on the harbours between Plymouth and Portsmouth, a reading that was actually too far to the west for Rutter and may perhaps have been an unintentional by-product of Yukon II, the rehearsal at Bridport on Lyme Bay on 23 June. The 3 Air Fleet summary of 23 June had located the main concentrations of 'vessels suitable for landing purposes' between Selsey Bill and Portland, and Brixham and Plymouth; and on 6 July aerial reconnaissance reported 50 LCAs between Salcombe and Start Point. Lyme Bay was well to the west of the Rutter and Jubilee embarkation ports between the Solent and Newhaven. For these and doubtless other reasons, Naval Group West identified the coast from Le Havre to Brest as the most threatened sector. But the way in which the Bay of the Seine got more than its share of reinforcements, such as the heavy railway battery 'Gneisenau' to protect the mouth of the Orne, underlines the difficulty and urgency of deceiving the Germans about the actual concentration areas for the invasion two years later, areas which were much the same as those identified by von Rundstedt.[40]

Dieppe was too far east to be included in the Bay of the Seine, which the Germans defined as extending from Barfleur to Fécamp, or for that matter to serve as a sally port for an overland drive on the U-boat bases in Brittany, since there would have been no point in landing on the far side of a river like the Seine. Dieppe, in fact, was not thought to offer much to attract enemy interest, except for the assortment of GAF radar and navigational equipment dotting the clifftops along that stretch of the Channel coast. The harbour had been difficult to enter since the British sank ships off the entrance in 1940; the drawbridge controlling access to the inner harbour was kept closed, leaving only the outer, tidal harbour for immediate use. Dieppe was not included in June as one of the ten most important ports to be protected by controlled minefields to be laid at the harbour entrance when a landing was imminent; and it ranked in priority not only behind Cherbourg, Brest, Le Havre and the U-boat bases but also behind Boulogne, Dunkirk and Ostend. Group B of 38 Minesweeping Flotilla was based there, but the port was seldom used by E-boats, having no pens to protect the Navy's only offensive force in the Channel. Nor was it a particularly important link in the Dunkirk-Boulogne-Le Havre-Cherbourg coastal convoy chain except as a secondary port occasionally used by slow convoys unable to cover the long haul between Boulogne and Le Havre under cover of darkness. In 1940, because of its size and the amount of salvage work to be done, it played a minor role in preparations for Sea Lion as an alternative harbour to the main 'Einsatzhäfen' of Le Havre, Boulogne, Dunkirk, Ostend, Antwerp and Rotterdam. It was not visited by Raeder during his tour of inspection in France in the first week of August.[41] Not surprisingly, therefore, the Navy was content to leave its defence to the Army.

The coastal convoy traffic between Rotterdam and the French Channel and Atlantic ports was an integral part of the German infrastructure in the West. Hitler was well aware of its importance, this being one of the reasons why he attached so much urgency to building up the defences of the Channel Islands. In good weather, scarcely a night passed without convoy movements up- and down-Channel, many of the westbound ships carrying bulky building materials for Organization Todt to build U-boat and E-boat pens and coastal fortifications. Ships and escorts caught in daylight outside the security of heavily defended ports like Boulogne and Le Havre were liable to suffer severely at the hands of the RAF. Every so often, blockade breakers, armed merchant raiders or other valuable ships had to be escorted to or from Bordeaux. Important convoys like these were planned as special operations, which meant extra minesweeping and at times the sailing of decoy convoys, and seldom included more than one large ship at a time. Schiff 23 (the *Stier*) took from 10 to 19 May to steam in stages from the Elbe to the Gironde at the cost of two T(orpedo)-boats and two minesweepers sunk. On the other hand, slow convoys were taking up to two months to travel from Cherbourg to Vlissingen in the short summer nights, clogging the harbours in between. On an average night, there was still a good deal of activity, much of it minesweeping to keep the swept channel – Weg Rosa, or Herz, after 1 July – clear, as is indicated by the total of 89 German vessels at sea between Ostend and Le Havre on 15/16 August.[42]

The protection of this traffic was only one of Ruge's responsibilities, others being offensive and defensive minelaying, minesweeping, and the defence of the coast against attack. In the spring of 1942, his Defence Divisions of minesweepers, patrol boats and escorts were being subjected to increasing strain, especially 2 Defence Division whose sector ran from the Scheldt to Cherbourg; for example, units withdrawn to the Baltic for Operation Barbarossa in 1941 had not been returned or replaced. Encounters with British radar-controlled and radar-equipped MGBs and MTBs off Cherbourg and in the Straits of Dover were becoming increasingly common.[43] Worse still, destroyers were starting to operate on the French side of the Channel, especially south of Boulogne. These small Hunt-class ships such as *Albrighton*, which was involved in a skirmish off Fécamp in April, were no match for German Fleet destroyers, but there were none of the latter to spare for Channel operations. Even the Möwe class T-boats, which could meet the Hunts on something like equal terms, put in only occasional appearances between Brest and Rotterdam, since they were urgently needed in the Bay of Biscay. The two T-boats escorting the *Stier*, *Iltis* and *Seeadler*, were sunk off Boulogne on 12/13 May as they passed through 2 Defence Division's sector on passage down-Channel. On 5 May, five British destroyers were (wrongly) reported to have exchanged fire with 10 R-boat Flotilla at Point 17a on Weg Rosa, not far off Dieppe; interestingly enough, the Germans ruled out the chance of a landing on this occasion because of the cliffs. On 19/20 June, the motorship *Turquoise* and her escorts were ambushed on their way from Le Havre to Cherbourg by a force of three of the fast new steam-gunboats (SGB) led by *Albrighton*. Engagements like these were almost invariably fought at very close quarters, as again at Dieppe on 18/19 August, and German crews usually acquitted themselves well by suddenly opening up a concentrated defensive fire with guns of all calibres, including anti-aircraft armament. In the confusion off Fécamp on 22/23 April, one German vessel made the mistake of trying to come alongside *Albrighton*.[44] A good part of the trouble, in the opinion of naval officers from Raeder down, was Hitler's decision to use all available aircraft, including those engaged in offensive minelaying off the English coast, on retaliatory bombing raids, thereby freeing British escorts and other coastal units for operations on the French side of the Channel. Ruge realized that his escorts desperately needed help in the shape of a German version of the SGB or, more realistically, a defensive mine-barrier some 20 to 25 miles out and roughly parallel to the French coast.[45]

A mine-barrier consisting of a series of flanking minefields in mid-Channel – Flankensperre – was one of the measures discussed after the St Nazaire raid in March 1942, measures that were remarkable if for no other reason than showing how little thought had apparently been given to coastal defences until then. Only in April, for example, were ports like St Malo and Calais closed by barriers day and night and other harbours at night by nets, cables, lighters or floating beams. At first Group West opposed the suggestion of a mine-barrier in the Channel comparable to the one off the west coast of Jutland. The whole scheme would have to be improvised, since mines were in short supply and the Channel presented special problems, what with its shallow water, strong tides and currents and shifting seabed. Moored mines were rendered ineffective on a rising tide by the strong 'dip' caused by the flow of water. No German moored mine had been designed for such conditions, except perhaps the EMG specially developed for Sea Lion, and those that could be pressed into use had never before been used operationally on such a scale. They had no disarming device in case they broke loose from their cables; and drifting mines would endanger German shipping moving along the swept channel inshore of the barrier. Also, since the minefields were expected to guard against incursions by shallow-draught MTBs and MGBs, they would have to be laid close to the surface on extra long cables – in some cases with entangling lines on the surface – which in turn would inhibit E-boat operations; gaps would have to be left to allow for a speedy escape after torpedo or minelaying operations off the English coast. There were no E-boat flotillas operating in the Channel in May but Naval Group West expected a flotilla to be moved to Boulogne early in June. The barrier, finally, would take months to lay and would provide as much deterrence as real protection because British air superiority would enable the Royal Navy to cut mines free or sweep a passage through the barrier undisturbed and probably undetected. Little thought was evidently given to laying ground mines inshore of the 5-fathom line, which would have forced the British to use some sort of 'mine-bumping' craft rather than conventional sweepers to clear the approaches to a landing beach.[46]

Misgivings about the barrier were overcome in late April or early May, although the directive from Naval Group West was not issued until 31 May. Improbably enough, all thought of launching Sea Lion had not been given up, since everything was to be done to ensure that in that eventuality the mines could be swept quickly. Possibly Operation Abercrombie on 21/22 April had some bearing on the decision to go ahead, since the barrier would not only shield the convoy-swept channels but also provide protection against seaborne landings. Be that as it may: the reconnaissance of 20–23 June abruptly transformed plans for laying the barrier from a rather relaxed undertaking into a matter of urgency, with priority given to defence against landing operations. The first four lays (Zaunkönig, Südsee, Brama and Putra) were started on 11/12 July, not in the narrowest part of the Channel where surface action was most intense but in the Bay of the Seine north of the line Barfleur–Cap d'Antifer. By the end of the new moon period on 22 July, nine minefields had been finished, the two most northerly (D1 and D2) in mid-Channel as far east as Dieppe.[47] Because of the perils of bringing them through the Straits, no fast minelayers from Norway could be used, so that minesweepers, R-boats and even the occasional T-boats that happened to be passing through were pressed into opportunistic service. D1 and D2 were laid by T-boats of 3 Torpedo Boat Flotilla, four of which arrived at Le Havre on 16 July on their way west to Biscay; although the boats came to within 21 nautical miles of the English coast and went about their work in strict formation by the stop-watch, the B-Dienst intercepted no signals to indicate that they had been plotted by British shore-based radar. When Jubilee took place, the minelaying was still in full swing in the Straits and north-west of Cherbourg, and the raiding force passed between two 'barriers' laid in July; by 20 September, 34 had been laid in 25 operations since 12 July, with 4,124 mines fairly evenly divided among EMC, EMG and KMA(K) types, plus obstructor floats and dummies. The system was not finally finished until mid-November, but before the end of the month three German ships had been sunk within 24 hours by mines set adrift during storms in October.

While there had been no known British losses, there had certainly been a marked decline in destroyer incursions from across the Channel in the last quarter of the year.[48]

The mine-barrier formed a passive component in the rudimentary early warning system patched together that summer. In theory at least, GAF evening reconnaissance would detect any unusual collection of ships and landing craft forming up off the south coast. In the case of Rutter, this would have been quite possible considering that the first units were to set sail as early as 1330 on the day before the assault. There was no reason why the progress of a cross-Channel expedition could not be followed by a shadowing aircraft. Specially-trained Ju 88 crews worked with E-boats at night by drawing anti-aircraft fire to disclose a convoy's position and then drowning out engine noise as the E-boats closed for a torpedo attack. Unlike British MTBs and MGBs, E-boats were not yet radar-equipped, although they were being fitted with radar search receivers (*Funkmessbeobachtungsgerät*).[49]

As the raiding force approached the French coast before dawn, it would then have been picked up by pre-alerted patrol boats (*Vorpostenboote*) or coast-watching radar. The patrol boats were mostly converted trawlers or drifters assigned to set billets outside the main ports when the weather was favourable for a landing or when aerial reconnaissance was hampered by poor visibility. The patrol boats worked in pairs and, like the E-boats, had no radar. Two boats had been patrolling off St Nazaire on 27/28 March, and five billets covered Le Havre, two to the west and three to the north-east. On 28 July, a pair outside Cherbourg were attacked by a destroyer, one being sunk and the other badly damaged.[50] Ruge thereupon decided to move his billets closer inshore and to rely more on the mine-barrier; and since there were nothing like enough boats to form an unbroken screen, he decided to make a virtue out of necessity by using minesweepers on sweeps and convoys on passage to supplement his offshore patrols. Convoy commanders were instructed that breaking up a landing took precedence over escort responsibilities. Both Abercrombie and Bristle, it should be pointed out, had featured a disruptive encounter between the landing force and German units – sweepers of 38 Minesweeping Flotilla and harbour protection boats respectively. The collision on 18/19 August was not entirely unforeseen, therefore, and Ruge claimed all the credit he could for the Navy, the more so because the Army was thought to have slighted the naval contribution to the defence of Dieppe.[51] There were no patrol boats based in Dieppe, although four boats of 14 UJ Flotilla had been there briefly in July, and on 18/19 August only two of the five Le Havre billets were occupied. Closer to shore, however, there were harbour protection boats, three of which (together with the pilot boat on its way out to meet the convoy) narrowly escaped being run over by the first wave of landing craft at Dieppe; these were really floating listening posts armed with a machine-gun and alarm rockets. At Dieppe, the boat on Position 36 and the signal station on the east headland simultaneously fired the first alarm rockets at 0435 when craft approaching Pourville failed to respond to its recognition challenge.[52]

Supposing this warning system had worked as intended, obviously the price of detection for a raid of any size would have varied according to strength of the objective. There were still plenty of open stretches of the French coast in 1942 where the overnight loss of tactical surprise would not necessarily have had calamitous results, although there was always the risk of an earlier than otherwise intervention by German reserves. For a frontal assault on a well-defended port like Boulogne or Le Havre, however, the cost would almost certainly already have been prohibitive. Even for a raid like Jubilee against a less heavily defended port, at the very least the assault craft would have been forced to run the gauntlet of a predicted barrage by coastal batteries, followed by concentrated fire from the divisional artillery and infantry weapons the instant the first landing craft were illuminated by flare or searchlight. Worst of all, though, would have been an E-boat ambush during the Channel crossing, such as Baillie-Grohman mentioned in a letter of 8 June to Leigh-Mallory. E-boats were a supremely offensive force and were ordered to the Channel in response to the doubling of the convoy traffic between the Thames and Portsmouth in May, but they also had definite anti-landing assignments and were usually ordered to immediate or half-hourly readiness by Naval Group West when weather conditions were considered favourable. On 28/29 August, for example, 5 E-boat Flotilla undertook an offensive recce – *Aufklärungsvorstoss* – north-east of Cherbourg in response to the threat of a landing in the Bay of the Seine that night.[53] In June, only one flotilla was operating in the Channel, but by 1 July there were two, and by 1 August three; on 3/4 August no fewer than 19 boats took part in an abortive torpedo operation in Lyme Bay. Their bases at Boulogne and Cherbourg were well-positioned for flank attacks on any force approaching the Bay of the Seine, or Dieppe.

Notably enough, the flotillas achieved their greatest success of the summer on 8/9 July, less than 48 hours after the cancellation of Rutter, when eight boats of 2 S Flotilla sank six ships in a WP (Bristol Channel–Portsmouth) convoy without loss to themselves; C-in-C Plymouth later ordered that no convoy was to be at sea between Land's End and Portsmouth at night – an interesting reversal of the precautions in force on the French side of the Channel – and the Admiralty were anxious to capture an E-boat, for they were superior in a number of ways to MTBs. The captured Jubilee Operation Orders drove it home to Führer des Schnellboote (F d S) what an incomparable target nine stationary LSIs lowering their landing craft inside the mine-barrier, or strung out as they passed through it, would have been for a night torpedo attack. Hughes-Hallett admitted that the LSIs were very vulnerable to a surface attack from the time they stopped to lower their LCAs until they returned (unescorted) to the English side of the Channel.[54] The failure of the escorting destroyers on 19 August to detect a convoy of five motorships and three escorts well within range of *Slazak*'s and *Brocklesby*'s radar implies at least that they would not have been equal to anticipating a well-prepared attack by a force of fast and elusive E-boats. Once the first star-shell was fired, the usual confused free-for-all would have ensued, made worse by the presence of a gaggle of gunboats, motor launches, landing craft, chasseurs and the like.

As it happened, the Germans broke up the concentration of 2, 4 and 5 S Flotillas in the Channel with which they started the month of August. The F d S had decided by the end of July that neither minelaying to block the British convoy passage through the Straits nor torpedo attacks further west were any longer profitable from Boulogne and Cherbourg respectively. Minelaying within range of British coast-watching radar was pointless because convoys were simply diverted until the mines were swept, and torpedo operations in the western part of the Channel were dependent on often undependable GAF reconnaissance to track the convoys. The recent appearance of the RAF at night had put an end to

Channel operations for four nights before and after the full moon. Having three flotillas of E-boats used as diversionary cover for laying a mine-barrier did not find favour, either, with F d S and his staff at Scheveningen. He wanted 2 and 4 Flotillas withdrawn to Dutch bases to prey on the heavily travelled convoy route between the Thames and the Humber. On 4 August, Naval Group West held out strongly against this proposal because of the danger of landings; an E-boat flotilla should be based at both Cherbourg and Boulogne until the middle of September. But by 16 August, one flotilla was back in Rotterdam and a second on its way to Ijmuiden. On the night of the greatest torpedo opportunity of the year, the four remaining five Flotilla boats based at Cherbourg were laying mines in Lyme Bay – the wrong kind of operation in the wrong place – while the four boats at Boulogne did not leave port, although they had been brought to half-hour readiness because of the favourable weather forecast. When it briefly seemed possible on the morning of 19 August that a large-scale landing was in its initial stage, the flotillas in Holland were ordered to stand by for a return to the Channel the following night.[55]

If a graph could be drawn to show the rise and fall of German anxiety about a major landing, the line would shoot up dramatically in the last week of June and rise even higher with the Order of 9 July and the start of a period of favourable tides (or so it was supposed) on 10 July. Such a level of intensity could not be maintained indefinitely, of course, and for much of the time for the rest of July the weather was unsuitable for putting to sea, regardless of moon and tide. After a period of favourable tides from 27 July to 3 August, the line would taper off quite noticeably from 4 to 9 August, only to creep up again with the timing of high tide from 10 August. Again, though, there were periods of high wind and rain, such as 10–14 August, during which naval authorities stopped, for example, bringing available E-boats to half-hourly readiness at sunset.

Military authorities tended to take less notice of the weather than their naval counterparts did and credited the enemy with the ability to land under any conditions; the Englishman was widely regarded as being equally at home in his fog as the Russian in his frozen wilderness. At the highest level, the General Staff were unimpressed by the threat of a Second Front; von Etzdorf, the Foreign Office representative at OKH, reported on 8 August that it had not escaped notice that across the Channel there were no indications of serious preparations for invasion – no sufficiently large concentrations of shipping, no ban on leave, no postal restrictions, no evidence of reinforced fighter defences or of offensive activity against GAF ground installations either, no troop concentrations. But it was a different story for German staffs in the West at the level of Army HQ and below, which had no choice but to react urgently to the PR results in June, reinforced by the Führer Order of 9 July, and had to contend with the very real prospect of raids. Haase of 15 Army – nothing if not a continentalist in strategic outlook, since he took the description of Jubilee as a combined operation to mean a combination of British, Canadian and US *troops* – surpassed himself in working his own troops up to the keenest possible pitch. On 3 July he referred to 3,000 'Sturmboote' across the Channel, each one armoured and capable of running up on shore; the British, he estimated, could put 80,000 troops ashore in the first wave. To the dismay of Naval Group West, Haase ordered on 7 July that all vessels approaching to within 3 km of the coast at night were to be fired upon without warning as from 2000 on 10 July. Troops manning the coastal defences, particularly those belonging to 321 ID between Wissant and the Somme, were already given to illuminating German ships by searchlight before challenging for the recognition signal. On 7 July, an Army battery at the mouth of the Somme opened fire – most improbably – on a submarine. As luck would have it, Haase's shoot-on-sight order coincided with the transfer of 400 invasion barges left over from 1940 from the Channel ports to Rotterdam. The 'Prähme' were to be moved at night inshore of the swept channel, taking advantage of High Water and other conditions that happened to coincide with the requirements for landing operations. Predictably enough, one of the first convoys of barges was riddled with anti-tank and machine-gun fire as it passed the mouth of the Somme on 12/13 July. On 15 June, incidentally, Haase and his staff had dismissed the growing number of agents' reports about a Second Front as so much nonsense being put about by the enemy to keep troops and commanders tense and on edge.[56]

This kind of 'unrest', as it was described to Raeder, could not be tolerated for long. The 15 Army order was cancelled and efforts made to improve the liaison between the Sea Commander

Seine-Somme and 321 ID. Besides, after six or seven weeks the novelty of the 'Sturmboote' began to wear off. There had been no noticeable redeployment of the landing craft, whose estimated strength remained fairly steady: 725 in the first week of August as against 728 in July. General Admiral Saalwächter, commander of Naval Group West, assured Raeder during his tour of inspection at the start of August that the lessons of St Nazaire had been applied and a landing could be awaited with a certain peace of mind – *mit grosser Ruhe* – despite the enemy's air superiority, technical superiority in radar, and growing strength in light coastal forces.[57] Ruge issued an order on 14 August referring to the *veränderte Gesamtlage*, meaning that things had generally taken a turn for the better thanks to the build-up of the Army in the West and the enemy's growing commitments elsewhere. He proposed to move the faster boats of 15 Patrol Boat Flotilla to Boulogne/Dunkirk now that their presence in the Bay of the Seine was 'superfluous'. The occupation of patrol billets there and along the north coast of Brittany was to take second priority to the escort of a procession of important ore-carrying ships – *Grängesberg*, *Fidelitas*, *XXIV Maggio* – up-Channel on their way back to Germany. The slower patrol boats were to serve as escorts for the even slower barges heading for the Rhine. Two days later, Captain Bramesfeld remarked on the end of another new moon period, bringing with it a decreasing threat. The possibility of a landing for the sake of propaganda still existed, most likely in the Bay of the Seine; it largely depended, according to Bramesfeld, on how much the British had been able to find out about the minebarrier and how far this would affect their plans.[58]

This abatement of German nervousness helps to explain the decision to redeploy the E-boats and the perceptible shift in naval attention in mid-August away from the Bay of the Seine eastward towards the Straits. With the diversion of 15 Patrol Boat Flotilla and 38 Minesweeper Flotilla, 2 Defence Division could no longer provide escorts for the stretch between Le Havre and Cherbourg unless 3 Defence Division to the west agreed to help out. The *Fidelitas*, said to be carrying mercury, sailed from Cherbourg to Le Havre on 14 August with a 3 Defence Division escort (UJ 1404 and 1411). On 17 August, the Italian ore-carrier *XXIV Maggio*, could not sail from Le Havre unless the two UJ escorts were cleared as far east as Boulogne.[59] They were thus setting up their fateful rendezvous with the Jubilee force on their return westward.

Finally, to return to the movements of III/KG 26. First, this was no ordinary bomber unit but rather the first torpedo bomber unit to operate in France and still the only one available there when it departed for Banak on 10 August; the course at Grosseto in June had been a torpedo conversion course. Its specialist crews had been misused in Baedeker raids on Birmingham, but it did launch the first torpedo attack by aircraft on a convoy in the Western Approaches on 3/4 August. Secondly, the mission to Norway turned out to be a wild goose chase. Although there were indications from GAF reconnaissance of Hvalfjord in Iceland that a convoy had sailed between 2 and 4 August, and although a U-boat reported it at sea on 6 August, there was no PQ sailing that month. The OIC noted on 10 August that ten U-boats off northern Norway had been pursuing a mythical PQ convoy since 5 August, thanks partly to bad weather which hampered aerial reconnaissance. U 405 contributed to the confusion by failing to respond to requests for confirmation for an exasperating number of days. Yet it was known at Naval Group North before 16 August that there was no prospect of a major convoy action for the time being. Therefore there was no need to detain the Ju 88s of III/KG 26 in Norway; but there they remained until they finally departed on 19 August, arriving back at Rennes the following day. Moreover, the bulk of two long-range bomber groups were moved from the West to central Germany en route for Russia three days before the raid.[60]

In short, not only did Dieppe find itself in a relatively unconsidered corner of the Channel coast, but key naval and air units were moved out of reach to Holland and Norway respectively a week to ten days before Jubilee. On land, on the other hand, the number of divisions in the West increased from an estimated 28 in the Martian Report of 14 July to 36 on 7 October. And so, the Army ended up with stronger reserves, including three Panzer divisions and the SS Panzer Corps of two divisions, than were necessary to repel any raids the Allies were likely to be able to launch, thanks almost entirely to one man's susceptibility to the threat of a Second Front.

NOTES

1. KTB, 1 Skl, Teil A, 22., 25., 27.6.42, T 1022 1673; KTB, MarGpW, 23., 24.6.42, T 1022 3974.
2. KTB, 1 Skl, Teil A, 1.5.42, T 1022 1673; KTB, Kommandant der Seeverteidigung Seine-Somme, 11.6.42, T 1022 2371. Oberbefehlshaber West (Oberkommando Heeresgruppe D), Fernschreiben, Ia Nr 1744/42 gKdos, 27.6.42, T 78 317. Seekriegsleitung, B Nr 1 Skl 15798, gKdos, 30.6.42 (Fernmündliche Rücksprache 1 Skl Ia mit FKpt Junge, OKW/WFSt), T 1022 1721. *KTB, OKW(WFSt)*, II (1), 23.6.42, p. 442.
3. ADM 199/421. Case 7392. Portsmouth War Diaries, Part I: Chronological Survey, 7 July 1942. KTB, 1 Sk1, Teil A, 7.7.42, T 1022 1674. WO 219/1933, Martian Report No. 7, 14 July 1942.
4. *KTB, OKW(WFSt)*, II (1), 25.6.42, p. 449. Trevor-Roper, *Hitler's War Directives*, Directive No. 40, pp. 171–7. *KTB, OKW(WFSt)*, II (1), 11.4.42, p. 319, 18.4.42, p. 324, 26.6.42, p. 451.
5. *KTB, OKW(WFSt)*, II (1), 25.6.42, p. 449, 26.6.42, p. 451, 29.6.42, p. 458, 12.8.42, p. 573. Trevor-Roper, *Hitler's War Directives*, Directive No. 45, pp. 193–7.
6. General der Pioniere und Festungen b ObdH, L III, gKdos, 'Aktennotiz über Führerbesprechung am 2.8.42 (21.30–23.30) im Führerhauptquartier', 3.8.42, T 78 317.
7. CAB 79/56, COS(42)46th Meeting (O), 27 May 1942. General der Pioniere und Festungen b ObdH, L III, gKdos, 'Aktennotiz über Führerbesprechung am 2.8.42 (21.30–23.30) im Führerhauptquartier', 3.8.42, T 78 317. WO 199/3008, 'Operation Sledgehammer – Appreciation of Army Forces Available', HF/00/144/G(Plans), 15 June 1942.
8. KTB, 1 Skl, Teil A, 4.5.42, T 1022 1673; *The Halder Diaries: the Private War Journal of Colonel General Franz Halder* (2 Vols, Boulder, CO, 1976), II, p. 325.
9. General der Pioniere und Festungen b ObdH, L III, gKdos, 'Aktennotiz über Führerbesprechung am 2.8.42 (21.30–23.30) im Führerhauptquartier', 3.8.42; General der Pioniere und Festungen b ObdH, L III, 'Niederschrift über die Besprechung beim Führer über den Atlantik-Wall am 13 August 1942 (21.40–00.50 Uhr)', 14.8.42; Oberkommando des Heeres, Op Abt (IIa), 'Führerrede zum Ausbau des Atlantik-Walles am 29.9', 3.10.42, T 78 317.
10. Otto Dietrich, *12 Jahre mit Hitler* (Munich, 1955), p. 154.

OKW/WFSt brought up the question of retaliation for British sabotage raids in the shape of German raids on the south coast of England, but the German Navy considered that the possible advantages were heavily outweighed by the risks (KTB, 1 Skl, Teil A, 27.4.42, T 1022 1672). KTB B, AOK 15: Ia 4874/42 gKdos, 21.8.42: 'Ob West rechnet damit, dass der Engländer aus Prestigegründen neue Unternehmen durchführt'. T 312 502.

11. *KTB, OKW(WFSt)*, II (1), 3.6.42, p. 400. FO 898/184, PWE Propaganda Research Section: Weekly Reports. M. Balfour, *Propaganda in War 1939–1945* (London, 1979), pp. 277 ff. John Mendelsohn (ed.), *Covert Warfare, No. 17: The German View of Cover and Deception* (New York, London, 1989), Colonel General Erhard F.J. Rau, Foreign Military Studies Manuscript P-044b, p. 9. There is no sign in the Martian Reports that this stratagem was successful even for a time between June and mid-November 1942. *Akten zur Deutschen Auswärtigen Politik 1918–1945, Serie E: 1941–1945* (Göttingen, 1974), III, 'Aufzeichnung des Gesandten Krümmer', 28.7.42, pp. 237–8.

12. See, for instance, Warlimont's 'Erläuterungen' to Greiner's notes in *KTB, OKW(WFSt)*, II (1), 14.8.42 (pp. 588–9), 4.9.42 (pp. 678–9), 8.9.42 (pp. 701–3).

13. 'Gegen . . . die Ansicht zum Abzug der mot Div "Gross-deutschland" nach dem Westen hatten der Chef des GenStdH und der Chef WFStab einhellig Einspruche erhoben . . .' *KTB, OKW(WFSt)*, II (1), 'Erläuterungen des Generals Warlimont . . .', 12.8.42, p. 576. The Grossdeutschland division had actually begun its move to the Abbeville area when Hitler was forced to give in to the pressure to keep it on Army Group Centre's front on 30 August (*KTB, OKW(WFSt)*, II (1), 30.8.42, p. 658).

14. Geoffrey Jukes, *Hitler's Stalingrad Decisions* (New York, 1985), p. 39: Jukes is mystified by Hitler's 'almost obsessive concern' (p. 16) with the threat of a major Allied landing in the West in the summer of 1942; he might have found Hitler's attitude less of a puzzle had he taken account of the PR results at the end of June, which he never mentions.

15. B.H. Liddell Hart, *The German Generals Talk* (New York, 1948), p. 231; Oberbefehlshaber West (Oberkommando Heeresgruppe D), Fernschreiben, Ia Nr 1744/42 gKdos, 27.6.42, T 78 317.

16. Oberbefehlshaber West (Oberkommando Heeresgruppe D), 'Grundlegender Befehl Nr 1', Ia Nr 1001/42 gKdos, 28.3.42, T 78 317;

KTB, 1 Sk1, Teil A, 15.6.42, T 1022 1673, contains a good summary of the report of 6 June and the reaction in evoked at OKH, OKW etc. KTB, MarGpW: 'Lagebetrachtung zur Frage Feindlandung im Westraum', B Nr gKdos 3217/42, 8.8.42, Anlage 5 zum KTB v 1.8–15.8.42, T 1022 3974. Oberbefehlshaber West (Oberkommando Heeresgruppe D), Fernschreiben, Ia Nr 2610/42 gKdos, 1.9.42, T 78 317.

17. See Warlimont's 'Erläuterungen' in *KTB, OKW(WFSt)*, II (1), p. 610. 'Bericht von Major i G Pipkorn über die Fahrt nach Paris und Arangere [?] vom 26–28 Juni 1942', 29.6.42, T 78 317. The categorization of operations into strategic, operative and tactical came naturally to all branches of the German Armed Forces – see, for example, Luftflottenkommando 3, 'Befehl des Luftflottenkommandos 3 für die Kampfführung in der Küstenverteidigung', Führ Abt/Ia op, Nr 4173/42 gKdos, 3.5.42, T 312 502. KTB, 1 Sk1, Teil CX, 'Niederschrift über Besichtigungsreise des ObdM nach Frankreich vom 3 bis 6 August 1942', 6.8.42, RM 7/226.

18. Martin Geyer, 'National Socialist Germany: the politics of information', in E.R. May (ed.), *Knowing One's Enemies* (Princeton, 1984), pp. 310–46. *Akten zur Deutschen Auswärtigen Politik, E*, III, pp. 75–6: Runderlass des Reichaussenministers von Ribbentrop, Multex Nr 467, 29.6.42. The three-volume collection of reports – 'AKTEN betreffend "Zweite Front"' are reproduced on T 120 165, including the undated 'Vormerk' that according to von Ribbentrop '. . . soll eine Weiterleitung dieser Meldungen an militärischen Stellen vorläufig nicht erfolgen'.

19. GenStdH, Abt Fremde Heere West, 'Die britische-amerikanischen Operationsmöglichkeiten gegen Europa und Afrika im Jahre 1942', Nr 632/42 gKdos Chefs, 1.3.42, T 78 450. Liss to General ?, 14.10.42, AL 1669: Liss referred to '. . . den alten Widersachern von der Abwehr . . .'

20. GenStdH, Abt Fremde Heere West III, Nr 761/42 gKdos Chefs, 13.6.42, AL 1455a: 'Anliegend wird auf fernmündliche Anforderung das einzige noch vorhandene Exemplar der auf Führerbefehl eingezogene Denkschrift . . .' [Nr 632/42 gKdos Chefs, 1.3.42]. Walter Warlimont, *Inside Hitler's Headquarters 1939–1945* (trans R.H. Barry, London, 1964), p. 280.

21. GenStdH, Abt Fremde Heere West, 'Lagebild in Britischen Reich', Nr 724/42 gKdos Chefs, 10.5.42, T 78 450; Generalstab des Heeres,

Abt Fremde Heere West, Nr 799/42 gKdos Chefs, 9.7.42, T 78 451. Liss's letter mentioning his visit to Dieppe (to 'Bernhard', 31.8.42) and other valuable letters are to be found in his Personal Correspondence File (AL 1669, IWM).
22. The troop convoys from the UK to the Middle East were designated WS for no other reason, apparently, than that these were the Prime Minister's first two initials (C.B.A. Behrens, *Merchant Shipping and the Demands of War* (London, 1955), p. 218); there is no correspondence between the convoy departure dates given by Behrens (p. 244) and the dates to be found scattered throughout KTB, 1 Sk1, Teil A, for the first half of 1942.
23. In February 1942, the Naval Staff (Operations) resurrected an estimate of May 1941 of the tonnage available to the British for an invasion of s-w Spain on the grounds that it was still valid. Assuming that it would take 51 ships of 255,000 tons to transport a division and that 306 ships (1,530,000 tons) could be made available for a brief time, it had been estimated that the best the British could do would be to land one division at a time over a period of months, the convoy that landed the first division returning in time to embark the seventh; it would take 3 months to land 16 divisions (KTB, 1 Sk1, Teil R, Anlagen zum Teil A, Seekriegsleitung, B Nr 1 Sk1 I op a, 3473/42 gKdos, 23.2.42). The contrast between this leisurely procedure and Torch is dramatic.
24. KTB, 1 Sk1, Teil A, 30.7.42, T 1022 1675. CAB 79/56, COS(42)12th Meeting (O), 21 March 1942.
25. Seekriegsleitung, 'Lagebetrachtung zur Frage "Feindlandung im Westraum,"' B Nr 1 Skl Ib, 1363/42 gKdos Chefs, 20.7.42, T 1022 1721. *KTB, OKW(WFSt)*, II (1), p. 701 (footnote). A report from Sweden on 9 July put 3 British and 2 US Army Corps with 240–250 tanks along the Caledonian Canal (KTB, 1 Skl, Teil A, 14.7.42, T 1022 1675).
26. WO 219/1933, Martian Report No. 2, 9 June 1942.
27. WO 219/1933, Martian Report No. 12, 19 Aug. 1942.
28. Seekriegsleitung, B Nr 1 Skl I op, 16934/42 gKdos, 11.7.42, T 1022 1721: '. . . eine umfassende Beurteilung der Landungsmöglichkeiten an den französischen Küsten . . .' This appreciation was followed by another on 22 July of landing possibilities in Holland – Seekriegsleitung, B Nr 1 Skl I op, 17843/42 gKdos, T 1022 1721. WO 106/4222, CC(42)108, 'The Selection of Assault Areas in a Major Operation in North West Europe'.

29. KTB, BSW: 'Zusammenfassender Überblick für die Zeit vom 16.5-31.5.42', M 444; KTB, 2 SichDiv: 'Lagebeurteilung am 1.8.42', T 1022 3552; KTB, MarGpW, 28.6.42, T 1022 3974.
30. KTB, MarGpW: 'Beurteilung Feindminentätigkeit in Zusammenhang mit Landungsoperationen (Ende März bis Anfang Mai 1942)', B Nr gKdos 2053/42 A 1, 9.5.42, Anlage 6 zum KTB v 1.5-15.5.42, T 1022 3974.
31. Seekriegsleitung, Fernschreiben, B Nr 1 Skl I op, 17153/42 gKdos, 14.7.42, T 1022 1721; KTB, 1 Skl, Teil A, 15.7.42, T 1022 1675. These were the timings for Operation Wetbob – CC(42)42(3rd Draft), 30 July 1942 – which referred to 'the ideal zero hour . . .' (RG 331, SHAEF, Office of Chief of Staff, Secretary, CC(42) Meetings to CC(42) Papers). KTB 302 ID, Ia Anlageheft 'Dieppe' zum Kriegstagebuch Nr 3: 'Bericht über den englischen Angriff beiderseits Dieppe am 19.8.42', Abt Ia, Tgb Nr 1105/42 g, 28.8.42, T 315 2019.
32. Marine Verbindungsoffizier zum OKH (GenStdH), B Nr 28/42 gKdos Chefs, 4.4.42; Oberkommando der Kriegsmarine, 1 Skl I op, 718/42 gKdos Chefs, 14.4.42, T 78 450.
33. KTB, 1 Skl, Teil A, 16.7.42, T 1022 1675. KTB, 302 ID, Ic Anlageheft E zum Kriegstagebuch Nr 3: Armee-Oberkommando 15, Ic/AO, 6.7.42; 302 ID, Abt Ic, 'Taktik englischer Landungsunternehmen', 338/42 g, 29.5.42; Anlage 5, Armeeoberkommando 15, Ic/AO 6.7.42, T 315 2019. Oberkommando der Kriegsmarine, 3 Abt Skl, FM Nachrichtenauswertungen Nr 23, 18.5.42, T 1022 1841. KTB, MarGpW: 'Feindnachrichten und Landungsabsichten', B Nr 2601/42 gKdos, 9.6.42, T 1022 3974.
34. The two prisoners captured during Bristle (3/4 June) had been instructed to claim that they had crossed from England in landing craft and not to mention the LSI *Prince Albert*; they evidently followed instructions and were believed (KTB, MarGpW, 4.6.42, T 1022 3974). See also AIR 16/758, CCO to C-in-C Portsmouth, 1802B/20, 20 May 1942. The OKW War Diary stated that 300–400 landing craft took part in Jubilee, protected by 13–15 cruisers and destroyers; there was also a floating reserve of 6 transports and 3 freighters and an 'operational reserve' of 6 transports (*KTB, OKW(WFSt)*, II (1), 20.8.42, p. 615). Oberkommando des Heeres, GenStdH, Abt Fremde Heere West III, 'Lagebericht West Nr 708', 15.8.42, Handakten Etzdorf betreffend Lageberichte West Nr 511–712.

35. Oberkommando der Kriegsmarine, 3 Abt Seekriegsleitung, B Nr FL 18440/42 g, 8.10.42, T 1022 2364. Von Rundstedt considered two of his 'Basic Orders' of particular importance: Grundlegender Befehl Nr 10, 'Die Kampfführung bei feindliche Landungsversuche', 2.7.42; and Grundlegender Befehl Nr 13, 'Voraussichtliche Kampfmittel des Feindes und ihr Bekämpfung', 21.7.42. CAB 98/22, Combined Report, Annex 9, pp. 170–72.
36. KTB, 1 Skl, Teil A, 5., 10., 16.4.42, T 1022 1672.
37. KTB, 1 Skl, Teil A, 1.7.42, T 1022 1674.
38. Marinegruppenkommando West, 'Bildaufklärung engl Südküste durch Luftfl 3', gKdos 3927/42 A 4, 4.9.42, T 1022 1721; KTB, 1 Skl, Teil A, 4.9.42, T 1022 1676. The reference to 'kleine Teile Raum Southampton . . .' was made by Ic Lufl 3 in a telephone call to Naval Group West (KTB, MarGpW, 1.8.42, T 1022 3974). AIR 24/562, Notes and Appreciations No. 45, 'Enemy High Altitude Raids on August 24 and 25 1942', Intelligence 3, HQ Fighter Command, 29 Aug. 1942; AIR 40/182, AOC-in-C Fighter Command to Undersecretary of State for Air, 14 Sept. 1942.
39. AIR 24/562, Notes and Appreciations No. 43, 'Reconnaissance Activity 3/122', Intelligence 3, HQ Fighter Command, 26 Aug. 1942. Luftflotte 3 ruled out a 'schlagartige Bilderkundung' of the English east coast on 6 September without even mentioning the Ju 86P, claiming that the recce over the south coast had been flown by an Me 109 (KTB des Kommandos der Marinestation der Nordsee, 8.9.42, T 1022 4026). KTB, 1 Skl, Teil CX, 'Niederschrift über Besichtigungsreise des ObdM nach Frankreich vom 3 bis 6 August 1942', 6.8.42, Vortrag Chef des Stabes Luftflotte 3 (Oberst iG Koller), RM 7/226. WO 219/1934, Martian Report No. 14, 2 Sept. 1942.
40. KTB, MarGpW: 'Feindnachrichten und Landungsabsichten', B Nr gKdos 2863/A 1, 26.6.42, Anlage 3 zum KTB v 16.6–30.6.42, T 1022 3974. KTB, 1 Skl, Teil A, 6.7.42, T 1022 1674. KTB, MarGpW, 26.6.42, T 1022 3974.
41. KTB, 302 ID, Ia Anlageheft C zum Kriegstagebuch Nr 3: 'Besprechung des Div-Kommandeures mit Hafenkommandant Dieppe am 4.4.42', T 315 2019. Peter Schenk, *Landung in England* (Berlin, 1987), p. 212.
42. KTB, 1 Skl, Teil A 13.5.42, T 1022 1673; KTB des Befehlshabers der Sicherung West, 15/16.8.42, M 444.
43. Oberkommando der Kriegsmarine, Seekriegsleitung, B Nr 1 Skl I

op 9099/42 gKdos, 20.4.42, T 1022 1721, gives a good account of the increase in British activity on the French side of the Channel.
44. KTB, MarGpW, 26.4.42, 5.5.42, 19.6.42, T 1022 3974; 26 British prisoners were taken after the sinking of the *Turquoise* when one of the new SGBs was sunk. Weg Rosa was renamed Weg Herz on 1 July. ADM 223/110, AIDAC 1050B/25, revealed that the German force on 23/24 April consisted of an R-boat depot ship and 8 R-boats on passage from Boulogne to Le Havre. AIDAC meant 'to be deciphered by Officer Only'.
45. *KTB, OKW(WFSt)*, II (1), 13.4.42, p. 321; KTB, BSW: 'Zusammenfassender Überblick für die Zeit vom 1.6–15.6.42', M 444.
46. KTB, MarGpW: 'Minenverwendung vor der Küste', B Nr gKdos 2819/42 A 3, 23.6.42, Anlage 1 zum KTB v 16.6–30.6.42, T 1022 3974.
47. Marinegruppenkommando West, 'Weisung zum Auslegen von Flankensperren im Kanal', B Nr gKdos 3035/42 A 1 Chefs, 31.5.42; KTB, MarGpW: 'Beurteilung Flankensperren', B Nr gKdos 4382/42 A 1, 5.10.42, Anlage 5 zum KTB v 1.10–15.10.42, T 1022 3974.
48. KTB, 1 Skl, Teil A, 3.7.42, T 1022 1674; KTB, MarGpW, 22.7.42, T 1022 3974; Marinegruppenkommando West, 'Beurteilung Flankensperren', B Nr gKdos 4382/42 A 1, 5.10.42, KTB, MarGpW, T 1022 3974; KTB, 1 Skl, Teil A, 26.11.42, T 1022 1677.
49. See, for example, the report of the success of 2 E Flotilla on 8/9 July (KTB, FdS, 9.7.42, T 1022 3253).
50. KTB, 2 SichDiv: 'Neue Weisung für Vorpostendienst im Kanalgebiet', FS gKdos vom BSW, 28.7.42, T 1022 3552; KTB, BSW: 'Zusammenfassender Überblick für die Zeit vom 16.7–31.7.42', M 444.
51. Naval Group West took particular exception to von Rundstedt's dismissive reference to the Navy – '. . . wenigstens etwas mit den Seestreitkräften geschehen würde . . .' in Oberbefehlshaber West (Oberkommando Heeresgruppe D), Ia Nr 201/42 gKdos Chefs, 17.9.42, T 312 502.
52. 'Bericht des Hafenkommandanten Dieppe über die Kampfhandlungen am 19.8.42', n.d., as forwarded by Marinebefehlshaber Kanalküste, B Nr G 17535 A 11, 30.8.42, T 1022 2364.
53. The small flags floating in the sea that were observed from the landing craft on their run in to the beaches were ranging flags for the two Army coastal batteries (2/770 and 813); in the event of an imminent landing, they would be brought to Action Stations

by Seekommandant Seine-Somme; there were no naval coastal batteries covering Dieppe. AIR 16/760, Baillie-Grohman to Lee-Mallory [sic], 8 June 1942. KTB des Führers der Schnellboote, 28.8.42, T 1022 3253.
54. KTB, MarGpW, 9.7.42, T 1022 3974. The date of this action is wrong in Roskill's Official History (*The War at Sea*, II, p. 163) and in Admiralty records (ADM 199/2112, Trade Division History, Vol. 34, p. 40), where it is given as 7 and 9/10 July respectively. KTB, FdS: 'Abschliessend', 30.8.42: 'Erfahrungen aus den erbeuteten Operationsbefehlen werden gezogen; sie weisen der S-Bootswaffe ausreichend Ziele in den vielen eingesetzten grossen Transportschiffen, die in wenigen Meilen Abstand vor der Küste ihre Boote aussetzten'. T 1022 3253.
55. KTB, FdS: 'Abschliessend', 31.7.42, 30.8.42, T 1022 3253. Marinegruppenkommando West, Fernschreiben, gKdos 3236, Chefs, 4.8.42, KTB MarGpW, T 1022 3974. KTB, FdS, 18., 19.8.42, T 1022 3253.
56. *Akten zur Deutschen Auswärtigen Politik, E*, III, Nr 172, Etzdorf an das AA, 88/42 gRs, 8.8.42, pp. 294–5. KTB B, AOK 15: Armeebefehl, 19.8.42; 'Fahrten durch das Armeegebiet', Nr 29/42, 3.7.42, T 312 502. KTB, 2 SichDiv, 8., 9.7.42, T 1022 3552. KTB C, AOK 15, Allgemeines: Ic Lagebericht für die Woche vom 8.6–14.6, 15.6.42, T 312 502.
57. KTB, 1 Skl, Teil CX, 'Niederschrift über Besichtigungsreise des ObdM nach Frankreich vom 3 bis 6 August 1942', 6.8.42, RM 7/226. Raeder reported on his tour of inspection – with satisfaction – to the Naval Staff in Berlin on 6 August (KTB, 1 Skl, Teil A, 6.8.42, T 1022 1675).
58. KTB, 2 SichDiv: 'Weisung BSW', F S gKdos 4855, 14.8.42; 'Lagebeurteilung am 16 August 1942', T 1022 3552.
59. KTB, 2 SichDiv: 'Geleitdienst', 14., 15., 16., 17., 18.8.42, T 1022 3552.
60. ADM 223/94, 'GAF Notes 12–18 July', OIC/SI 268, n.d.; 'GAF Notes 2–8 August', OIC/SI 292, 9 Aug. 1942; 'U/Boat situation, week ending 10.8.42', OIC/SI 297, n.d. KTB, Marinegruppenkommando Nord, Anlagen 23 u 24 zum KTB vom 15–31.8.42, RM 35 I/143. ADM 223/95, 'GAF Notes 16–23 August 42', OIC/SI 316, 23 Aug. 1942.

4

Radar and the Raid

The threat to seaborne assaults posed by radar in a coast-watching role is self-evident. By 1942, radar had become a decisive factor in all naval and air operations in the Channel area. At the start of 1941, the British had located only nine German radar stations but by year's end that figure had risen to over 70. The best planned and executed of all Combined Operations raids was the one in February 1942 to remove parts of a GAF Würzburg apparatus at Bruneval. Afterwards, R.V. Jones (then ADI [Science] at the Air Ministry and the chief British expert on German radar) received a stream of calls from planners at COHQ 'who wished to make their mark by planning a successful raid. I forget how many radar stations were earmarked for raids in this way, although there was relatively little point in that we had found most of what we wanted to know at Bruneval'.[1] Both Bristle and Barricade had a connection with radar. Oddly enough, though, the role of German (and British) radar during the raid on Dieppe has never been satisfactorily explained in published accounts.

German publications tend to concentrate on GAF radar and its primary responsibility of aircraft tracking and reporting, and mention Jubilee only in passing. For example, Karl Otto Hoffmann's *Geschichte der Luftnachrichten Truppe* (1962) is a two-volume work crammed with detail but containing some basic mistakes in the short passage on Dieppe. Hoffmann has two GAF Freya equipments in Dieppe at the time of the raid – F 28 north-east of the harbour and F 223 on the west headland. In fact there was only the one Freya – F 28 on the west headland – with a Würzburg (W 223) close beside it. A second Würzburg was in position roughly where Hoffmann put F 28 but this was a gun-ranging equipment (TC 911) at the disposal of Flakuntergruppe Dieppe. In another general survey (*Die Radarschlacht* (1977)), Werner Niehaus begins his account of the raid unreassuringly: '*Am 18 August 1942, ein warmer schöner Donnerstag* . . .' The day of the raid was actually a Wednesday.[2] Although both publications are emphatic that F 28 detected the Jubilee force, neither offered any hints as to the arrangements under which GAF radar might have been operated for surface surveillance at night.

Among publications in English, James Leasor's *Green Beach* (1977) and Jack Nissen's *Winning the Radar War* (1987) both cover the abortive attempt by the former 'RDF expert' to remove parts of F 28 during the operation. *Green Beach* is backed up by an impressive list of authorities consulted but is otherwise undocumented; and confidence in its general reliability is undermined by over-dramatized passages whose accuracy fails the test of double-checking against primary sources. A case in point is Leasor's version of the collision between Group 5 carrying No 3 Commando to the most easterly beach at Berneval and the coastal convoy on its way south from Boulogne. Contrary to Leasor, Oberleutnant z S Wurmbach, the convoy commander, did not fire a recognition signal, for the very good reason that he knew he could not have come across another German convoy. When UJ 1404 fired a star-shell at 0348 to illuminate the grey shapes of the enemy, he most certainly did not turn to shout 'one hoarse, dreadful word to his radio operator: "Invasion!"' Far from it: Wurmbach failed to appreciate that he had crossed courses with an amphibious force and put the encounter down to a by no means unusual ambush by MGBs and MTBs. True, his aerial was shot away soon after the shooting started but Leasor is mistaken in claiming that Wurmbach had no other means of communicating with shore. He could always have fired the pre-arranged rocket signal to alert the naval signal station in Dieppe that a landing was in progress – had he but realized it before 0523, when he claimed to have rammed a large 'Sturmboot' carrying an estimated 80 men.[3] *Green Beach* also confused the timing of events leading up to the landing by failing to distinguish clearly enough between British Summer Time (BST) and Deutsche Sommerzeit (DZT), the latter being an hour ahead of BST and two hours ahead of Greenwich Mean

Time (GMT); there is even a mention of Double BST, which had ended on 8 August.[4] Nissen's subsequent memoir also has trouble with the convoy incident, confusing the two German anti-submarine trawlers with E-boats and misspelling *Slazak*, not to mention the more serious blunder of describing the coastal batteries at Berneval as 'radar-directed'. When he witnessed the spectacular display of star-shell and tracer from the landing craft carrying him to the beach at Pourville, Nissen was convinced that the expedition had been picked up by radar. 'With my knowledge of British radar', he wrote, 'I knew that any force attacking the coast of Britain would be detected at least twenty miles out to sea.'[5]

Nissen was content not to pursue the implications of his observation. F 28 dominated the landing beaches on either flank of its position on the west headland, and would seem to have posed a serious threat of detection. What in fact were the technical capabilities of German radar in the Dieppe sector in August 1942? How accurate was British intelligence about F 28 and W 223? While there are admirable publications in English dealing with wartime scientific and technological intelligence, such as R.V. Jones's *Most Secret War* (1978), none of them answers such specialized questions as these; and information in documentary sources is often unreliable and vague. The Rutter documents refer to a (non-existent) RDF station at Berneval; and the Jubilee Operation Order claims that 'a second RDF (Freyer) [*sic*]' had been located on the cliff-edge approximately 300 yards 'west of the RDF station at 207682' and lists 'RDF stations' as an objective to be captured, along with the municipal gasworks, post office and so on. But nothing is said about radar's operational functions. The subject of radar was not included in COHQ's Questionnaire on Jubilee submitted to the Force Commanders on 25 August, and the Combined Report compiled in October 1942 limited itself to another reference to a 'Freyer' station.[6] The planning papers for cross-Channel operations in 1943, by comparison, contain a good deal of information, including recognition drawings of the various German equipments. Presumably less was known about German radar in 1942 than later in the war and what information did exist was protected by tighter security. There is also reason to believe that some of the relevant files were thoughtlessly destroyed, for the weeding of files by the Ministry of Defence before their presentation at the PRO has apparently not always been conducted in the most professional way. R.V. Jones expressed his dismay in 1979 at finding that only half a dozen or so of over 500 of his files had escaped destruction. Jones, incidentally, served on the Low Cover RDF Committee, which, improbably enough, was chaired by Hughes-Hallett for a time in 1941–42. It was characteristic of the latter that he should claim credit for having suggested dropping Jones and another young scientist by parachute in the Bruneval raid, a rash and eccentric idea that Sir Henry Tizard quickly stamped out. Jones's own version of this episode never mentioned Hughes-Hallett; far from being angry and upset at the cancellation, he was 'decidedly relieved'.[7]

Coast-watching radar along the coasts of Occupied Europe was actually the responsibility of the German Navy, not the GAF. Thanks to a good selection of German naval war diaries and files, it is possible to piece together a fairly complete picture of the construction of a chain of shore stations designed to operate under the direction of regional plotting centres. The basic equipment for use against surface craft and gun-ranging was the Seetakt, which so closely resembled the GAF's Freya in physical size and appearance that only the trained eye of a PR interpreter could pick out the difference in aerial array and therefore wavelength – 80 cm for the Seetakt and 2.5 metres for the Freya.[8] The British first intercepted transmissions on the Seetakt frequency of 370 MHz in October 1940 and connected them with gun-ranging for the heavy batteries opposite Dover and Folkestone; from the spring of 1941, the Royal Navy successfully subjected them to interference. In the latter part of 1941, the Germans set up their first plotting centre (*Ortungszentrale* or *Ozet*) at Boulogne with responsibility for the coastal waters between the Scheldt and Le Havre. But the rate of delivery of Seetakt equipments from the manufacturer was well below the Navy's expectations and requirements. At the time of the raid on St Nazaire, there were nine Seetakts in operation along the west coast of France, with plans for doubling that number under the control of four Ozets at Cherbourg, Brest, Lorient and La Rochelle; but when Raeder carried out his tour of inspection early in August, none of those Ozets had yet gone into operation.[9]

From Ruge's point of view as Commander Western Defences an efficient coastal radar chain offered a number of advantages.

Radar had great potential for navigation, because if a satisfactory Identification Friend or Foe (IFF) device could be developed, there would be no need to hand the British coastal forces an advantage by burning lights along the coast. Such an IFF beacon was prepared on an experimental basis for the Channel Dash of the *Scharnhorst, Gneisenau,* and *Prinz Eugen* in February 1942, but failed to work properly. The Germans had still no operational IFF when the Dieppe raid took place, which explains why the lights at the harbour entrance and the lighthouse at Pointe d'Ailly were showing on 19 August. An important measure of a radar system's effectiveness was its power to neutralize the enemy's system by jamming or the use of spurious reflectors or to resist jamming itself. During the Channel Dash, German jamming enjoyed considerable success, having been gradually stepped up in advance so as to preserve surprise. The German battlecruisers had been within range of British shore radar a full two hours under cover of ten GAF and four Navy jammers (*Störsender*) before the B-Dienst intercepted Dover's first W/T transmission of a plot at 1245.[10] When an attempt was made to repeat their success later in 1942, however, the results were disappointing. The loss of *Seeadler* and *Iltis* on 12/13 May was blamed on failure to jam British radar. Then, after weeks of careful preparation and planning, 18 naval and 12 GAF jammers were turned on at midnight on 21 June under the control of the naval station at Mont Lambert, near Boulogne. Ten minutes later, E-boats of 2 and 4 Flotillas crept out from behind the shelter of Gris Nez in pursuit of two Channel convoys. Barely 15 minutes later came the first B-Dienst intercept of a Dover plot of the E-boats, signifying a German defeat in this electronic trial of strength. British jamming in the meantime had obliterated the plots of the CE (Portsmouth–Thames) convoy at Dungeness and the CW (Thames–Portsmouth) convoy just south of Dover.[11] By his own admission it was worth more to Ruge to jam British radar than to operate his own system free from interference. Success on 21/22 June would have meant the resumption of E-boat operations in the Straits and some much needed protection for convoys moving at night between Dunkirk and Boulogne. Thus superiority in radar would have gone far toward offsetting German naval weakness in the Channel and North Sea. Instead, the experience underlined what Ruge had been saying for months about the inferiority of the German naval radar system in terms of the range and precision of its equipment, the number of its stations and its resistance to jamming.[12]

It would not, however, be true to say that the naval chain was totally unequal to its chief task, that of guarding against surprise landings on the coast. Technical innovation could quickly turn the tables in this kind of contest. The new Lorenz coast-watching equipments (410 MHz) being introduced in the Calais–Boulogne area in July and August were expected to resist jamming better than the Seetakt. Moreover, the range of the jamming was much the same as the range of detection, German jamming of CD/CHL not being effective beyond the stretch from North Foreland to Beachy Head; therefore, once away from the Straits, jamming from England was not a serious factor, and any sudden attempt at it from ships or aircraft before a raid would have been so unprecedented as in itself to amount to the surrender of surprise.[13] The first combined operation when a Seetakt might have come into play was at St Nazaire. None of the published accounts of Operation Chariot says anything about preserving surprise by evading detection by radar. Yet, according to the records of Naval Group West, there was a Seetakt well placed by the naval signal station at Chemoulin, just west of the port, to detect the destroyer *Campbeltown* on the final stage of her approach up the Loire. This equipment was attached to a naval coastal battery and was meant to do double duty against both surface and aerial targets. During an air raid, the operator was expected to check the sea approaches now and again; but at the critical time on 28 March – between 0028 and 0215 – he evidently failed to do so sufficiently often while the equipment was being used in support of a Flak battery engaging bombers through cloud cover. As a result, the *Campbeltown* was first spotted visually at 0215 by the battery look-out.[14] The presence of a radar installation on the Loire estuary was known to COHQ as early as November 1941, when there was apparently some question as to whether it was capable of detecting surface vessels. The diversionary bombing of St Nazaire was meant to cover the approach of the Chariot force but has generally been considered a failure, since clouds prevented the dropping of bombs, while the noise of the aircraft aroused the town's defences. In any case, even if the diversionary requirements had included radar, success would have been accidental, because the operator would presumably have detected the *Campbeltown* had he followed standing orders to the letter. Hughes-Hallett attributed surprise not to diversionary air activity but rather to the spring tides that made it possible to approach St Nazaire over mud-flats in the Loire estuary.[15]

On the comparatively few occasions thereafter in 1942 when a hostile target large enough to be plotted came within range of stations on the French coast, German radar played the same sort of ill-defined role. The German ships that came into contact with *Albrighton* on 23 April were forewarned by shore-based radar.[16] Operation Bristle on 3/4 June was a small-scale raid compared to Jubilee but large enough to require the participation of the LSI *Prince Albert*. *Prince Albert* and four accompanying MGBs followed astern a CE convoy to a turn-off point south-east of Dungeness. They were not reported during the day by GAF reconnaissance. During the night approach to the French coast between Boulogne and Le Touquet, the two Boulogne radar stations that were operating paid close attention to the CE convoy. *Prince Albert* lowered the landing craft ten miles offshore and at once returned to Portsmouth. The main objective of the 250-man landing party was to create a gap in GAF radar coverage and bring back 'certain important parts of RDF apparatus'. The apparatus in question was located at Plage St Cecily, where there was a double Freya station within 60 yards of the beach and a Würzburg some 600 yards further inland. A pre-arranged signal would mean that both Freyas had been either dismantled or put out of action for at least several hours, information that was urgently required by the fighter force taking part in the planned offensive sweep the following morning while covering the withdrawal of the LCAs and MGBs. As it turned out, Bristle was partly undone by the navigational difficulties that were the bane of raiding forces at this stage of the war; the seven LCAs spent almost an hour cruising up and down the coast until they found the right beach, by which time it was too late to achieve the main objective. Both sides claimed in their reports that the enemy had turned tail.[17]

Not the least remarkable feature of Operation Bristle was the fact that the Germans had no inkling of the presence of a ship the size of the *Prince Albert*. Not only was *Prince Albert* not plotted but the carefully rehearsed story repeated by British POWs that they had made their way across the Channel in landing craft was believed by their interrogators. German documents reveal, however, that their coast-watching radar was not entirely neutralized by what, according to Group West, was the first British use of jamming for offensive purposes. Naval stations successfully plotted targets moving along the coast on a southerly and then south-westerly course. Judging by their speed, these were taken to be destroyers – quite correctly. This covering force was followed from 2335 to 0218 and the last station to report was the GAF Würzburg (W 223) in Dieppe. The closest Seetakt to the landing beach at Cap d'Alprech was being subjected to jamming, while the GAF equipments at St Cecily were preoccupied by low-flying aircraft at the time of the landing; but for that, according to Naval Group West, the Würzburg could have plotted the landing craft. On the opposite side of the Channel, the Commodore Commanding Dover, who had overall command of Bristle, drew his own conclusion on 12 June: 'It must not be assumed that such a vessel as the *Prince Albert* could escape detection on another similar occasion.' It was obviously a matter of urgency for Group West to close the remaining gaps in the Seetakt chain, because there was no naval radar between Cap d'Alprech and Le Tréport, south of the Somme estuary, and none between there and Fécamp, west of Dieppe. The obvious place for the second of these stations was on top of the cliffs at Pointe d'Ailly, just west of Dieppe, where the height of the equipment above sea level would maximize range.[18]

A Seetakt became operational on Pointe d'Ailly on 19 July, earlier than promised by Group West, so that the equipment could help with the current mine-laying operations. Another station plugged the gap north of the Somme at Berck Plage, while two more were added in the Bay of the Seine, at Arromanches and Pointe de la Percée. Pointe d'Ailly performed well from the start and was not caught up in British counter-measures further up-Channel, which on 19 August affected to varying degrees all Seetakts from Le Tréport to Wenduyne. During the mine-lays in the third week of July, five M-class minesweepers and four T-boats operated out of Le Havre. On two successive nights – 20/21 and 21/22 July – they laid mines in mid-Channel between Fécamp and Dieppe. On the first night Fécamp and Pointe d'Ailly plotted the M-boats but not the T-boats; but Pointe d'Ailly tracked both groups the following night, the T-boats to a range of 18 nautical miles on their outward passage. On 5/6 August, Pointe d'Ailly successfully plotted the convoy protecting the aircraft catapult ship *Schwabenland* moving on an easterly course along Weg Herz. On paper at least, the range of the Seetakt for large ships was from 30 to 40 km and for destroyers from 20 to 30 km, suggesting that Pointe d'Ailly stood an excellent chance of detecting eight stationary LSIs lowering their landing craft 16 km offshore or, failing that, the Hunt class destroyers as they closed the beaches in the wake of the landing craft. The war diarist of 2 Defence Division considered detection by radar in such circumstances a virtual certainty – '*mit sicherheitsgrenzender Wahrscheinlichkeit anzunehmen*'.[19]

Unfortunately, from the German point of view, Group West moved the Pointe d'Ailly equipment away about a week before 18/19 August without bothering to let 2 Defence Division know. Pointe d'Ailly resumed operation in September, so that its two initial periods of operational activity straddled the date of Jubilee.[20] The decision to move it elsewhere might be taken as yet another indication of the comparatively low priority assigned to the Dieppe area since Berck Plage and the two new stations in the Bay of the Seine remained in place. By landing at Dieppe at this particular time, the Allies took advantage of the reappearance of a 'blind angle' in coast-watching coverage between Fécamp and Le Tréport. Le Tréport played a minor, inconclusive part in the radar events of the early hours of 19 August, and neither it nor Fécamp had the range to plot the vessels standing off Dieppe while the operation was in progress. Was the 'blind angle' a piece of opportune good luck from the British point of view? Or was it an example of the application of very accurate current intelligence?

The answer is luck. The planners at COHQ had no idea that a Seetakt had been installed at Pointe d'Ailly, let alone that one had been set up after Rutter but removed before Jubilee. Thanks to the survival of R.V. Jones's 'Interim Report on German Coast Watching RDF Stations' of 29 July 1942, it is possible to check the most up-to-date intelligence on the subject. Apparently, suspicions about a German chain of coast-watching stations were of quite recent origin. The first Seetakt station to be identified as such was Fécamp in March 1942. Fécamp was too close to Dieppe to be the next station in the Freya chain, and scrutiny of a PR photograph revealed that its aerial array was narrower than the Freya's 2.5 m. A PR search turned up evidence of a chain being linked from Fécamp to Borkum. The equipment at the Hook of Holland was 'teed up' on a tower, as Jones put it, to maximize range and was photographed from 150 feet on 24 May. By the end of July, seven more had been identified, including a pair near Boulogne (Mont Lambert and Cap d'Alprech), but of course there was no mention of Pointe d'Ailly which had only just gone into operation. Other stations were known to exist without having been pinpointed, but only one was listed between Boulogne and Fécamp, appearing simply as 'Cayeux' in the Appendix, without a frequency reading. Later on, it was photographed on 17 August, although as Le Tréport it had been in business for many weeks, including the period covering the dates for the launching of Rutter. Jones asked for a high priority for sending out investigative flights and for keeping a look-out for 370 and 470 MHz transmissions; but although 18 coast-watching stations had been pin-pointed by November, only after the turn of the year did a station on Pointe d'Ailly show up on British lists as 'Varengeville'.[21]

It was not the case, though, that the absence of a naval coast-watching station necessarily meant that no surface surveillance was possible on any given stretch of coast. Following the raids on Bruneval and St Nazaire, local arrangements were evidently made to improvise by making GAF radar available in a coast-watching role at night. By the summer of 1942 the GAF aircraft reporting chain had achieved a high level of coverage, accuracy and completeness of detection against RAF Fighter Command's daylight offensive sweeps. Extending from Trondheim to Biarritz, the chain consisted of Freyas and Würzburgs, with the Freyas spaced at roughly 100 km intervals along the coast with a Würzburg alongside and others at intervals in between; nine of the 18 GAF stations identified by 20 June were immediately on the coast and five of the rest were no more than a quarter of a mile inland. The Freya was an early warning equipment with a range of up to 300 km, depending on the altitude of the target, and usually accurate to within one km in range and one degree in bearing. Freyas on the French coast were known to be capable of plotting aircraft well over southern England, which was why Leigh-Mallory was careful not to draw attention to the Yukon exercises in June by flying high cover sorties in strength. In practice, however, Freyas seldom reported plots beyond 90 km. When they did report, they provided range and bearing and some estimate of numbers but not the altitude of the approaching formation. The plots of several Freyas were passed to GAF Regional Control which had prior notice of the movement of all GAF aircraft within its area; once a track was identified as hostile, it was allotted to the Freya in the best position to track it.

The Freyas worked on the vertical beam principle, with a width of about 15 degrees, some rotating continuously while others reciprocated through an arc. The attached Würzburgs had a narrower beam of about ten degrees and a range of about 40 km and depended on the Freya to be put on to a target; they measured altitude by calculating range and the angle of elevation of the reflector disc required to produce the target's maximum response on the receiver. The chain's main function was to report the bearing, range, altitude, time of observation and approximate strength of an approaching formation to the local aircraft reporting centre (*Flugmeldezentrale* or FMZ) and fighter control. There was no theoretical reason why both types could not detect surface vessels but the scope of their activity in this role would of course be narrowed by their communications net. In the Dieppe sector there was a certain amount of air activity at night but no nightfighter units, limiting F 28 to working exclusively with the dayfighters of Jagdgeschwader 2. The local FMZ was in the Golf Hotel on the west headland and three supporting Würzburgs provided overlapping coverage: W 179 just south of the Somme estuary, W 223 alongside F 28, and W 167 to the west of Dieppe at Les Petites Dalles.[22]

Thanks to its superb cliff-top location, F 28 was credited with a range of 30 km against large surface targets. There was, of course, no question of reading shipping across the Channel, which is over 110 km wide between Dieppe and Beachy Head; nor for that matter were surface vessels moving along the coast south of the Somme normally within range of British shore-based radar. In the broader reaches of the Channel the Germans were dependent on preliminary GAF reconnaissance for any warning of the sailing of a hostile force on a course that could lead to its arrival off the French coast at dawn the following day. On the evening of 23 April, for example, 3 Air Fleet reported two destroyers and 25 'S-boote' on a southerly course off Portland and, since there was no opportunity to shadow them from the air, permission was given for GAF radar to be used in an overnight surface search. In the early hours of 24 April, targets were plotted off Fécamp and tracked as they moved east along the coast past Dieppe. One was in fact the destroyer *Albrighton*, which had led a dash at full speed across the Channel to intercept a convoy. The FMZ in Dieppe provided a running commentary and all three Regiments of 302 ID were ordered to Action Stations (*Gefechtsbereitschaft*).[23]

Quite by coincidence, a discussion had taken place on 22 April involving the Chief of Staff of 302 ID, Oberstleutnant i G Sartorius, the Fliegerhorstkommandant Dieppe, and Oberleutnant Weber, the commander of 23 (schwer) Flugmeldekompanie in Dieppe, about the implications of Bruneval and St Nazaire for the possible use of GAF radar against surface targets. F 28 and its 3 Würzburgs fitted almost exactly into the 70 km coastal sector held by the division, the next Freya station westward in the chain being the double one at Cap d'Antifer (F 10 and 11) guarding Le Havre. At what seems to have been their first meeting for this purpose, the officers agreed that the Würzburgs could be used against surface targets at night once a special request had been cleared by Jagdführer (Jafü) 2. The following day, possibly as a result of the previous night's excitement, it was agreed by telephone that the special request to Jafü 2 could be dispensed with so that the division could ask Weber directly for a surface search '*von Fall zu Fall*'. On 27 April, 3 Air Fleet decreed that GAF stations could be directed against surface targets when they came within range provided this did not interfere with an aircraft plot. Freyas were to be used beyond the Würzburg's range but only when they had no aircraft search or tracking responsibilities. Reports of shipping plots were to be sent to the FMZ and then the Ozet, where the plot would be evaluated and any necessary action taken to alert the Defence Divisions and Sea Commanders (*Seekommandanten*). There were three Sea Commanders on the Channel coast – Pas de Calais, Seine-Somme and Normandy – who were subordinate to the Naval

Commander Channel Coast (*Marinebefehlshaber Kanalküste*). West of Le Havre, where there were no Ozets to date, GAF surface plots were to be reported directly to Sea Commander Normandy.[24]

The Sea Commanders were the key figures in the Navy's system of coastal defence; they commanded, for instance, all naval coastal batteries, and Army coastal batteries were tactically subordinate to them. On paper at least, an offshore target reported by a naval signal station, battery observation post, FMZ, or other source would be identified as friendly or hostile by the Sea Commander with the help of the Defence Division concerned and Ozet. If the activity was hostile, the Sea Commander would then brief the coastal ID in that sector and order '*Feuer frei!*' for the batteries within range. For want of a reliable IFF, only naval authorities could tell what German ships were at sea and when they were liable to be seen, heard or plotted by watchers on shore. In practice, however, the Sea Commander could find himself at the centre of an agitated controversy about the identity of a possible target, a process of '*Ruckfragen hin und her*' that inevitably slowed up the response of the coastal guns. The presence of both German and enemy units was always a headache, as in the confusion off Fécamp on 23/24 April, when the Sea Commander Seine-Somme managed to prevent the firing of a defensive barrage. But the worst headache by far was the possibility of an FMZ passing on a GAF plot of what were in fact German ships directly to local, and probably jittery, Army authorities, leading to the opening of fire by the divisional artillery of one of the coastal IDs. This happened more than once. Across the Channel, the same problems were handled far more efficiently. Arrangements had been worked out whereby all stations on surface watch reported aircraft plots to the appropriate RAF centre, while stations on aircraft watch reported surface plots through a CD/CHL station to the Navy. British naval plotting rooms such as the one at Portsmouth monitored the positions of all shipping and hostile aircraft (some of which might be engaged in minelaying) in the Command, and passed on filtered reports to Army and RAF operational authorities. The Army, in fact, still had its own anti-invasion radar and plotting rooms in 1942, whereas the German Army was totally dependent on the other two services for any sort of early warning.[25]

Unfortunately for all concerned, GAF radar soon earned a reputation for eccentricity, whether because of inexperienced personnel or the poor location of some stations for surface plotting. On 19/20 June, for example, the motorship *Turquoise* (M 3800) was ambushed en route from Le Havre to Cherbourg by *Albrighton* and three of the fast new Steam Gunboats (SGB). The only German radar in a position to warn of the presence of the British force was a Freya at Grandcamp, but its efforts were unavailing and merely added to the urgency of setting up a Seetakt at Pointe de la Percée. Sometimes the GAF issued false alarms, one of which on 16/17 July achieved some notoriety when the Sea Commander Normandy reported, on the strength of a GAF plot, that a landing force was approaching Le Havre, causing a major commotion as patrol vessels were readied for sea in poor weather conditions. Soon it was discovered that the GAF had confused a cloudbank with an amphibious force. On another occasion, a GAF attempt to track a convoy from Le Havre to Cherbourg produced a plot overland across the Cotentin peninsula.[26] When a Seetakt was in a position to search an area where the GAF was reporting a surface plot, all too often the target was not confirmed, which was doubtless among the reasons why naval war diarists took to flagging GAF plots as such – 'Luftwaffengeräteortungen'.

An occasional touch of farce should not be allowed to conceal the fact that the improvised system could well have worked had Rutter been launched. Neither the Freya nor the Würzburg had been jammed or spoofed before June 1942, or indeed was to be subjected to technical interference for some time after Jubilee. One of Nissen's objectives was to discover what anti-jamming devices might have been built into the Freya design. In plotting surface vessels, the Freya's inability to determine altitude probably meant that the speed of a target, whether cloudbank or landing force, was the determining factor. But the Würzburg could get accurate bearings on ships by using horizontal polarization, and the three Würzburgs in the neighbourhood of Dieppe often joined in to plot convoys moving along the coast on Weg Herz. W 172, for instance, provided apparent confirmation of the agent's report of a landing on 12/13 June when it detected an elusive target off the mouth of the Somme. After rather a confused exchange with the FMZ in Dieppe and a check with the Ozet in Boulogne, 302 ID ordered the batteries south of the estuary to fire a defensive barrage. The naval chain was not always successful in plotting the movements of R-boats, yet on 9/10 July GAF radar plotted R-boats for half an hour as they entered Dieppe harbour.[27]

The Naval and Air Force Commanders for Rutter were aware of the connection between preliminary GAF reconnaissance and the risk of a subsequent radar search. Admiral Baillie-Grohman wrote to Leigh-Mallory on 8 June to express his concern about the possibility that an enemy aircraft might shadow the expedition across the Channel in the light June nights. The result could be an E-boat attack during the crossing and the loss of surprise at Blue and Green beaches. Leigh-Mallory replied that there was little he could promise to do about shadowers flying under radar cover except send out a few Hurricane intruders. 'If the night is very favourable, and they work round to the south, they [the Hurricanes] might manage to spot one of them against the northern lights, but I certainly cannot promise immunity against German reconnaissance.' So as to prevent 'German RDF' from getting any clear indication that Dieppe was the objective, Leigh-Mallory proposed to bomb Boulogne and Abbeville and Crécy aerodromes.[28]

At the meeting on 5 June when the bombing of Dieppe itself was dropped from the plan, he pointed out that the movement of aircraft between 0230 and 0400 would preoccupy the 'RDF organization' in Dieppe and so 'screen the final approach of our ships'. Seventy bombers were to raid Boulogne between 0200 and 0400, taking care to fly via Beachy Head and the Somme estuary at an altitude of at least 5,000 feet so as to be sure of being plotted. On 3 July, Leigh-Mallory switched the 40 sorties against Abbeville to the waterfront at Dieppe from 0345 to 0405. There was also a proposal for two motor launches, each towing a balloon with a radar reflector, to steer for Cap d'Alprech and stop five miles off Boulogne at about 0220. The 'masking' of German radar in the Boulogne area was not totally effective in a westerly direction, so they were likely to be detected; MGBs operating on main engines would add to the effect.[29] Leigh-Mallory's proposal acknowledged that GAF radar might be keeping an emergency watch for surface vessels; and since the equipments could not be jammed, they were to be distracted by the careful routeing of aircraft flying in indirect support of Rutter. But when Rutter was resurrected as Jubilee, there was still no provision for jamming Freya and Würzburg equipments, which makes it all the more surprising that any attempt to distract German radar was dropped from the plan. Now Abbeville was to be bombed by the US 8 Air Force to coincide with the start of the withdrawal from the beaches; the order of the day (or rather night) for the Jubilee assault was to be evasion.

The underlying reason for relying on evasion rather than distraction was that the nights were longer in August than in June, enabling the expedition to cross the Channel under cover of darkness. 'It is therefore highly probable', wrote Leigh-Mallory, 'that surprise can be achieved. This in turn will make it unnecessary to carry out any diversions, either by bombing or naval craft.' Hughes-Hallett was satisfied that success 'was chiefly due to the extreme care taken to avoid any suspicious movement of assault shipping or LSIs before the actual moment of sailing'.[30] The tactic of evasion apparently succeeded on 19 August, so that there was no post-operational second-guessing and therefore no documentation generated. MTBs and MGBs staged what turned out to be rather a futile demonstration off Boulogne, without balloons. Did Leigh-Mallory assume that German radar would not be operating unless there had been a preliminary alert from 3 Air Fleet? Or did the Jubilee planners take it for granted that a 24-hour watch was kept but that the LSIs and destroyers would be out of range and the landing craft too small to offer a target anyway?

Intelligence about the primary responsibilities and capabilities of the GAF early-warning chain was excellent. Both the Freya and Würzburg had been adumbrated in the Oslo Report in 1939 even before their transmissions were picked up from across the Channel. Gris Nez, Cherbourg and Dieppe were the sources of the first Freya transmissions fixed by D/F in 1940. The first PR photograph of a Freya was of the equipment at Auderville at the tip of the Cherbourg peninsula in February 1941. F 28 was photographed soon afterwards: 'Dieppe Caude Cote: 2.5 m 1000 pps Single equipment, and possibly 53 cm. Photographed.' The '53 cm' referred to W 223. The pieces taken from the Würzburg (W 110) at Bruneval in February 1942 provided Jones with 'a first-hand knowledge' of German radar technology.[31]

Other intelligence came from time to time from the GAF Enigma but, as Jones pointed out in a comprehensive report on the Freya and Würzburg, ' "DT" (German RDF) Freya and Würzburg,' of 10 January 1942, the best source of intelligence about the Freya's performance had been the practice of broadcasting plots on low-power W/T. In the beginning, the Germans broadcast plots of all aircraft but for obvious reasons soon confined themselves to hostile aircraft only; oddly enough, they continued this practice long after secure landlines had been made available connecting the radar station with the FMZ, fighter control, Flak batteries and the like, thereby providing the British with an extension of their own coverage. The Auderville Freya had no fewer than 15 cables leading from it, but it continued to broadcast plots of hostile aircraft. In May 1941 the practice was standardized for Holland, Belgium and France, and Jones found it 'memorable and informative' to monitor the data on which the GAF based interception tactics. 'Since the beginning of 1941', he wrote, 'it has therefore been possible to watch the main chain in operation, and ask questions of the enemy by sending out test flights and waiting for him to report them.' Not only had F 28 been photographed but it was also known that it could detect an aircraft at 1,500 feet at a range of 55 miles. Indeed, at the end of September 1941 the Dieppe Freya was found to be suffering from a systematic D/F error of from 10 to 20 degrees. Another German station, site previously unknown, revealed its position by plotting an RAF PR aircraft whose course was known from its photographs. The British gave this W/T traffic the codename Oculist.[32] German fighters, of course, were scrambled by landline while maintaining R/T silence for tactical reasons.

The RAF was naturally interested above all in the primary role of GAF radar, which was aircraft reporting; the main purpose behind the Bruneval raid had been to learn more about nightfighter control (GCI). A different matter altogether was the emergency use of GAF equipment against surface targets at night. In 1940 the destroyer *Delight* had been sunk by dive-bombers some 25 miles southwest of Portland Bill. According to Jones, GAF radar in the shape of the Freya on Cap de la Hague (Auderville) had been instrumental in this attack. 'As *Delight* had never been nearer than 60 miles to the station and had neither escort nor balloons, the claim might appear incredible if it were not fully authenticated.' In short, this was an exceptional and rather freakish incident. By 1942, however, the British Army Y service had been taking Oculist traffic for many months and among the intercepts were some shipping plots. It was certainly known at the latest by the end of July that the Freya could detect large ships at 'optical range' (15–25 miles) and that the Würzburg could provide accurate horizontal bearings on surface targets.[33] This intelligence was of obvious significance for combined operations but whether it reached COHQ cannot be confirmed.

Confidence in the feasibility of evading F 28's rotating beam may have been bolstered by the indifferent performance of comparable British equipment – Coast Defence/Chain Home Low Flying (CD/CHL) – against small surface vessels. The CD/CHL stations at Swanage and Portland, for example, did not perform very well in tracking the Yukon II force from the Isle of Wight to Bridport, especially when vessels moved inshore.[34] It was also known that the Freya was likely to be affected by the 'ground clutter' associated with vertical polarization. Whatever the case, any such confidence was misplaced. According to German sources, F 28 detected the Jubilee force as early as 0232; at about 0330, the stationary – and therefore surface – target began to move towards the coast in what looked like five columns. Those times are consistent with the timetable for lowering landing craft from the LSIs and forming them up for the run-in to the beaches. The sources in question are undeniably tenuous, being almost entirely derived from the GAF unit concerned. Mordal read Weber's 1959 article and was given a 'circumstantial account' of the night's activities by Hoffmann, who served under Weber and enjoyed access to the 'archives' of the Luftgau Nachrichten Regiment. Only one easily missed clue has been found in other documentary sources to confirm that the matter came to the attention of C-in-C West.[35] But one clue is enough, and there is no reason to doubt Weber's account of F 28's initial interception at 0232, thereby, incidentally, prompting answers to the two questions asked earlier.

First of all, the Jubilee force was plotted just over an hour before the collision with the convoy, not as a consequence of it. There had been no special request for surveillance from 302 ID or a preliminary alert based on an agent's report or sightings by reconnaissance aircraft. Therefore, if Leigh-Mallory did indeed assume that the operation of GAF radar in a coast-watching role at night depended on a search order inspired by an advance warning or by suspicious noises at sea, he was mistaken.

Secondly, as Hughes-Hallett's memoirs suggest, he at least appears to have assumed that German radar would be operating at night as a matter of course. There is some support for this point of view in Martian Report No 5, where a Freya station was estimated to require a party of thirty – ten by day and twenty by night. Hughes-Hallett took the credit for having moved the lowering-point for LSIs in Channel operations to at least 10–12 miles offshore, instead of only four as hitherto. This change in the spring of 1942 had serious navigational drawbacks because at that distance the LSIs would find it almost impossible to obtain a fix by identifying features on the coast, and the landing craft were liable to be carried up to six miles off-course by tidal currents during the two to three hours it would now take them to form up and close the beach. But Hughes-Hallett 'felt strongly that it would be rash to assume that German coastal radar was inferior to our own', and therefore the navigational problems should be accepted as the price of evading detection.[36]

It is still difficult to understand why diversionary bombing was dropped from the plan. Once the premise of a 24-hour radar search was accepted, the requirements for achieving surprise in such circumstances were surely the same as those for an operation that had previously been reported by aerial reconnaissance or agent's report. Also, Hughes-Hallett was no technical expert, despite his previous tenure as chairman of the Low Cover RDF Committee, whereas Jack Nissen was, and Nissen fully expected the Jubilee force to be detected en route. Not only was he aware that a German expedition would be located as soon as it came within 20 miles of the English coast, but a scientific friend on the staff of the Telecommunications Research Establishment warned him of the dangers of Jubilee: '. . . they will know [from radar] you are coming'.[37] The planners of operations against the Cherbourg peninsula certainly paid more attention to the question of German radar. Blazing called for an approach to Alderney from the north-west so as to stay out of range of radar on the mainland for as long as possible. In Wetbob's case, the channel chosen for sweeping through the new minefield off Cherbourg was to be as far away as possible from radar stations at Barfleur, Ver and Cap d'Antifer.[38] But for Jubilee, the planners apparently assumed that the LSIs and destroyers would escape detection at 10–12 miles and that the landing craft, some of which were wooden, would not be picked up at all.

So far, so good. What happened after 0330 is a little difficult to follow. According to Weber and Hoffmann, whose version was taken up and repeated with minor variations by Mordal, Niehaus and Leasor, the duty officer at the FMZ in the Golf Hotel had no success in having F 28's plot taken seriously by his naval contact, who maintained that it was either false or that of the convoy from Boulogne waiting to enter Dieppe on the tide. The duty officer then rang his CO, Weber, in his billet across the harbour at Puits, but Weber had no luck either in impressing the Navy. Weber himself claimed then to have called the 302 divisional HQ and spoken to Sartorius, who ordered Increased Vigilance (erhöhte Aufmerksamkeit), for the division, leading straight to catastrophic results for the landing force. This timely and decisive warning truly represented a great success for the 'Radarfunker', and led to the undoing of Jubilee. Mordal put Weber's call at 'a little after 0400' and struck a slightly discordant note by observing that the transmission of the 'general alert' was rather slow.[39]

Slow, quite frankly, was hardly the word for it. More significant, though, than its speed was the nature of the 'general alert', because the appropriate order in face of an impending landing was unquestionably Action Stations. Action Stations could come into effect as the result of an order from a higher HQ, an 'Alarmsignal', or a surprise attack. At Dieppe it was first ordered at 0501 by IR 571, not divisional HQ. 'Increased Vigilance' had nothing like the force of 'Action Stations', being more of an adjuration than anything else. The chronology of events (*Zeittafel*) included in the divisional report ('Bericht über den englischen Angriff beiderseits Dieppe am 19.8.1942', 302 Infanterie-Division, Abt Ia Tgb Nr 1105/42 g, 25.8.42) opens with reports of the convoy engagement off Berneval at 0347 but omits any order for 'erhöhte Aufmerksamkeit', which would surely have been included if Weber's call had had the dramatic effect he claimed. In the text that follows, 'Increased

Vigilance' is mentioned but almost casually in the context of the *Spannungszeit*, or general tension. The first mention of the FMZ in the chronology comes at 0458, when divisional HQ rang up to clarify the situation at sea. Whoever answered began reassuringly to the effect that a convoy was about to enter harbour; then, just as he was about to hang up, small-arms fire was heard from the direction of Pourville.[40] Fire had actually been opened at Berneval at 0450 and the naval signal station in Dieppe had shot off alarm rockets 15 minutes earlier. In other words, among the last posts to grasp that a landing was taking place were the FMZ and divisional HQ. The defenders of the town of Dieppe sprang into action with all the precision of ants in an ant-hill stirred with a stick – but that stick was not wielded by the 'Radarfunker'.[41] And the ants showed remarkable powers of recovery.

Several points are raised by this 'upshot' interpretation of how the alarm was raised. First, evasion of German radar failed but not at the cost of losing tactical surprise. Hughes-Hallett mistakenly assumed that the two factors were incompatible. Secondly, the collision with the convoy could be interpreted as a stroke of good fortune for Jubilee, although it has invariably been held up as an example of the sort of mischance that can ruin the best of operational plans. After all, if the Ozet had been in a position to confirm that there were no German ships in the vicinity of F 28's plot, then there would have been no reason for doubt and uncertainty to creep in. The presence of the convoy actually provided a form of cover in a way that the planners could never have anticipated, far less contrived. Thirdly, the outbreak of fire seems to have confirmed the Navy's reading of the situation. How else is it possible to explain the FMZ's initial response to the call at 0458? Besides, the order for Action Stations required something less than the certainty of an attack. All three Regiments of 302 ID spent the following night (19/20 August) in a state of *Gefechtsbereitschaft*, and the coastal divisions had all been at Action Stations for several successive nights in May and on occasion in June and July. Perhaps Weber and the duty officer at FMZ were less confident about F 28's plot at the time than they later made out. The War Diary of the Sea Commander Seine-Somme lists an entry from Artillery Regiment 527 at 0446 to the effect that enemy surface forces had attacked a convoy seven sea miles from the harbour at

0400; and that at 0430 the convoy was 14 to 17 km north of Dieppe and under constant attack by British 'Schnellboote'. The authority quoted for this information was the 'Ortungsgerät', which could only mean F 28. Perhaps, too, Weber assumed that the entire division was in the same state of readiness as the troops he commanded at Puits; together with the rest of Captain Schnösenberg's command from the East Mole of the harbour to Puits, they had been on a night exercise and did not stand down before dawn because of the gunfire at sea. What Weber and Hoffmann were both determined to stake out retrospectively was their unit's share of the credit for the successful defence of Dieppe – above and beyond, that is, their participation as infantry in the defence of Puits and the west headland.[42]

Had the duty officer at the FMZ followed the prescribed procedure, he would not have called 2 Defence Division or some local naval representative in town, as variously suggested by Mordal, Hoffmann and Leasor.[43] Instead, he would have telephoned the Ozet in Boulogne, as indeed he apparently did inasmuch as it was the Ozet officer of the watch who was rebuked by von Rundstedt for failing to react to the plot of 'Luftwaffengeräte' between 0252 and 0316. The Ozet itself had been under the command of 2 Defence Division until 27 July, when it was transferred to 2 Radar Detachment (*Funkmessabteilung*), whose War Diary for 18/19 August mentions no specific call from Dieppe, although it meticulously records the early progress of Convoy 2437 as tracked by the Seetakt stations at Cap d'Alprech and Berck. At 0021, a routine navigational warning was broadcast to guard against collision with Convoy 2412 northbound on Weg Herz from Dieppe. Just over 25 minutes later, Berck's plot of 2437 faded. The northbound convoy had been followed by Le Tréport from 2134 to 2313, and at 0243 the Ozet instructed Le Tréport to try to locate the southbound convoy. Coming 11 minutes after Weber's time for the initial contact and just one minute after an anonymous report of targets in BF 3387, a quadrant north-west of Dieppe, this search order may well have been in response to an unacknowledged report from the FMZ in Dieppe. In any case, Le Tréport soon ran into technical difficulties in measuring range and at 0307 its plot faded after locating the convoy briefly, far to the west of the swept channel.

DIEPPE REVISITED: A DOCUMENTARY INVESTIGATION

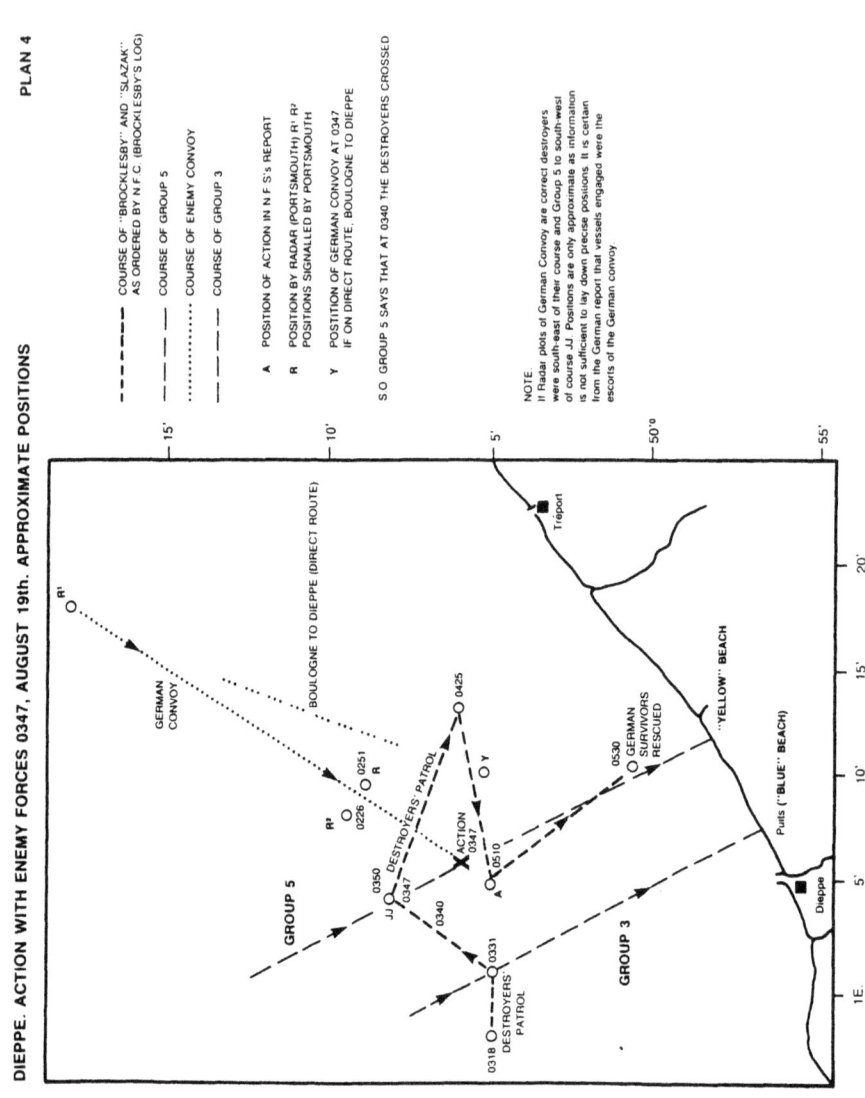

After that, it plotted scattered targets some 10 sea miles northwest of Dieppe, some of which were taken to be aircraft after engine noises and flares were reported in the general vicinity. Once it was learned at the Ozet at 0400 that the convoy was under attack, Le Tréport could still make next to nothing of its plot, and only at 0457 was it determined that the survivors of Convoy 2437 were setting course for that harbour. From the Ozet officer-of-the-watch's point of view, there was no need for further action since the collision had 'alarmed the coast'. Anyway, further action simply meant passing on a warning to the Sea Commander Seine-Somme, who was known to have already been directly in touch with the Seetakt station at Le Tréport (but who failed to get in touch with the batteries at Dieppe until two hours after the action started).[44]

If German authorities conducted an inquiry into the part played by radar in the early hours of 19 August, it was of a perfunctory nature, although steps were taken by the Sea Commander Normandy to improve the exchange of radar information by army and naval authorities within his sector. Nor is there any sign that the reputation of the GAF 'Radarfunker' as coast watchers was in the least enhanced by this episode, which may partly explain whatever hopes Weber and Hoffmann may have had for postwar vindication. On 11/12 December 1942, GAF radar in Dieppe provided warning of the approach of enemy destroyers from the north-west. It took Pointe d'Ailly 20 minutes to confirm this plot – too late to save two 'Sperrbrecher' making for Dieppe. If the Ozet had acted on the GAF plot, the German ships might still have had a chance of escape, but 'according to past experience' the GAF was not considered a reliable source.[45] The naval practice of flagging GAF plots continued into 1943.

The Dieppe Freya was naturally heavily involved in the air battle on 19 August, which gradually increased in intensity during the morning and reached a peak when German bombers began to operate in force around 1000. It broadcast aircraft plots on W/T but only after several hours, according to Hinsley. An expert from the RAF Section at GC&CS conducted an experiment in the War Room at 11 Group to find out what could be done to exploit various forms of Y intelligence for current tactical application during Jubilee. Of the eight separate categories, Oculist traffic was subsequently ranked last in usefulness, since it merely broadcast the known positions of RAF formations. It is slightly puzzling why the recording of the signals had to be 'rushed back to Mountbatten's headquarters', according to K. Dearson, who was apparently in charge of equipment on a ship lying off Dieppe. Dearson is now Nissen's backer for a VC. In *Green Beach*, which never mentioned Dearson at all, a German-speaking WAAF made recordings in a caravan on the Sussex coast. The GC&CS expert, on the other hand, simply listed the Oculist traffic under Cheadle, the main RAF interception station.[46] Presumably the signals revealed something about the upper handling capacity of a Freya during a major air battle; otherwise, their chief tactical use would have been to measure the effect of jamming. The Germans, of course, never broadcast the most important plot on 19 August – that of the stationary target located at 0232.

Significantly enough, the only COHQ post mortem into the role of radar during Jubilee dealt with the performance of British radar, and it was undertaken only at the instigation of the Canadian Army in the person of General H.D.G. Crerar. Crerar asked the Recorder of Combined Operations a few weeks after Jubilee why the German convoy had not been plotted and a warning of its presence sent to the Naval Force Commander. Mountbatten, who enjoyed a reputation as a wireless and signals specialist, replied on 5 October. An investigation had found that a small convoy had been detected off Boulogne at 2300 moving south down the coast. It was about 10 miles north and slightly east of Dieppe at 0300. The Admiralty was satisfied that those vessels had nothing to do with the force encountered at 0350 because the latter were closer inshore and beyond the range at which shore-based radar in England could detect vessels of that size. Three warnings about the first convoy were sent to *Calpe* at 0143, 0258 and 0334. Crerar accepted this reply, the substance of which found its way into the COHQ Combined Report on Jubilee in October, since he was in no position to appreciate how unintentionally misleading it was.[47]

Hughes-Hallett referred elsewhere to a 'tanker and escort' entering Dieppe harbour at 0330 as the vessels he assumed had been plotted at 0300. These entirely imaginary vessels had supposedly crossed Group 5's course 10 miles ahead of it. Confusingly enough, the press reported that the collision causing heavy Canadian casualties had been with a 'tanker and escort'. After reading the newspapers, it occurred to J.D. Cockroft, Chief Superintendent of the Air Defence Research and Development Establishment (ADRDE) at Malvern, that radar stations along the south-east coast should have been plotting German shipping in the Channel and that what was done with this information was of some importance for the future of combined operations. On his own initiative, he started an investigation whose results were to prove to be more accurate than CCO's, although not conclusive.[48] According to the ADRDE staff, three enemy vessels were plotted at 2140 on 18 August on passage south from Boulogne and followed until the plot faded south of Etaples at 2345. The stations reporting were K 148 and K 7, which were in touch with Dover and Newhaven. K 148 at Dover and K 7 just east of Hastings were part of the highly secret 10 cm CD chain that had been completed along the south-east coast early in July. The Type 271 equipments gave complete coverage 'from buoys to battleships' in the Straits of Dover; unlike the CD/CHL stations, they were not subject to jamming. The next station to report was Ventnor on the Isle of Wight (K 86), which identified an enemy force moving south towards Dieppe soon after 0200 and more or less coming to a stop not long before 0300. This was a very long-range plot even for the Mark IV High Power 271 equipment at Ventnor and was achieved only because of anomalous propagation, or unusual atmospheric conditions that made it possible to plot a target at exceptional range but subject to intermittent fading. Cockroft's results could not accommodate a tanker and escort. The Naval Staff Battle Summary for Dieppe (1959), released at the PRO in 1990, shows that the first warning to *Calpe* was transmitted by Portsmouth at 0127 and drew attention to 'small craft patrolling approximately 350 degrees Tréport 15 miles at 0100'. So far, this force had given no indication of entering the operational area. When later radar signals, which must have been picked up by Ventnor, suggested such a possibility, a second warning was sent to the Naval Force Commander at 0244: 'Two craft 302' Tréport 10 miles, course 190' 13 knots at 0226.' This signal was hardly reason to 'cancel the operation', as has been suggested. According to the Operation Orders, there could be no cancellation after the LSIs entered the swept channels through the minefield, and by 0244 the first of the LSIs had already reached the lowering point for landing craft.

Thereafter, senior officers of groups were to take drastic avoiding action if enemy forces were met on passage; Commander Wyburd of 5 Group, for his part, had determined in advance to continue his course and try to fight his way through. In short, there was no third signal and the vessels collided with were those plotted by Ventnor between 0200 and 0300. Neither signal appears to have reached Hughes-Hallett in *Calpe* or the captains of *Brocklesby* and *Slazak*, who continued to patrol on a predetermined course instead of altering to intercept the enemy vessels. Cockroft sent a copy of his findings to Professor Bernal (COHQ) on 11 November but, in keeping with the general lack of interest shown at Richmond Terrace in radar as a factor during Jubilee, it was never acknowledged. A note on the file cover dated 24/12 (1942) reads: 'Does Prof Bernal not intend to acknowledge this letter?' It was answered on 10 June 1944: 'Too late now!'[50]

There is sometimes an ahistorical temptation to ascribe standards of technical reliability and accuracy to equipment in its early days that were actually reached only later. It would be easy to make this mistake where radar in 1942 was concerned, because the performance of various pieces of equipment on 18/19 August belied any assumptions about their matching modern standards of efficiency. The British, for example, experimented with radar navigational aids, none of which worked very satisfactorily. Two MLs equipped with Type 78 D/F beacons were positioned to mark the entrances to the two swept channels through the mine barrier, but none of the Types 286 or 290 sets on listening watch on the destroyers and LSIs picked up their signals. Hughes-Hallett sponsored the first operational use of the RAF navigational device GEE for surface navigation. But use of 'RDF receiver QH' on various vessels had mixed results, although small craft were successfully using GEE to operate off the French coast by the end of August. The Naval Type 272 sets on *Brocklesby* and *Slazak* failed to detect the German convoy, although the pair of destroyers were close enough to have crossed astern of Group 5 at 0340. *Garth* intercepted two 'large RDF echoes' at 0328 at a range of 24 miles and bearing 260 degrees, but they turned out to be false. The ASV radar used by patrolling aircraft of Coastal Command drew a blank, as it had by mischance during the Channel Dash.[51] Still, British shore-based radar did surpass itself by providing more than an hour's warning of unidentified craft less than two miles from Group 5's line of advance; and a matching claim could be made for F 28, the more so because of its improvised role. Ventnor's warning was negated by a failure of wireless communication between Portsmouth and *Calpe*, and F 28's by a cumbersome system of inter-service liaison. Radar, in short, was an unpredictable factor in 1942 and therefore dangerous, like the proverbial accident waiting to happen, particularly for operations that were at all out of the ordinary.

To return, finally, to the question of surprise. No operation launched from the south coast after 23 June could count on strategic surprise. Rutter had a chance if it had been launched as originally scheduled on 20/21 June, but only because the Germans took so long to discover what was taking place in May and the first half of June. Strategic surprise, after all, was not so much lost as deliberately forfeited, given the desirability of presenting as much visual evidence as possible of preparations for a Second Front. When, for example, the new Combined Operations base opened on the Isle of Wight in May, the press published references to the security arrangements imposed on travel and communications between the island and the mainland. This publicity seemed to trouble Lord Swinton, Chairman of the Security Executive, but not Mountbatten. 'Since we cannot conceal from the Germans', he explained to Swinton, 'that we are intending operations, and, indeed, part of the object of these operations is to keep them in a state of suspense, I welcome any publicity that might arise from this scheme.' This exchange is yet another example of the tension between mounting a notional strategic threat and launching actual operations heavily dependent on tactical surprise, a tension only too readily exploited by Anthony Cave Brown and others. Swinton, at any rate, decided that tactical surprise would have to be safeguarded by imposing a ban three days before an operation was due to take place and randomly when none was pending.[52]

Leigh-Mallory acknowledged in his letter of 9 June that it was all the more essential in view of the lack of strategic surprise 'to gain as great a measure of tactical surprise as possible'. Thus there was some awareness on his part of a special obligation to guard against the loss of tactical surprise for raids. Yet Leigh-Mallory's idea of an acceptable measure of surprise fell markedly short of total. He was quite prepared to admit that GAF reconnaissance might discover that an expedition was on its way across the

Channel; but this risk was offset, in his opinion, by the fact that the enemy would not know where between Gris Nez and Le Havre it was going to come ashore. Tactical surprise had thus been narrowed to keeping the enemy guessing about the exact beaches chosen for the landing and the method and weight of the assault. Leigh-Mallory was more concerned about the movement of German reserves than about breaking through the coastal crust. There are signs in his letter of planning for as much surprise as possible and then hoping for the best in the shape of defences too weak to profit by forewarning. Even the false report of 13 E-boats at Fécamp on 8 June – which was almost certainly merely an example of the difficulty PR had in differentiating between E-boats and R-boats – was mentioned quite casually – '. . . I am keeping my eye on them'. Yet close support squadrons would not be available to provide protection against E-boats until they had returned to England and rearmed after fulfilling their missions during the assault; in other words, not before 0630 at the earliest. Hughes-Hallett's chief lesson after Jubilee was that it had been shown that it was 'still possible to achieve tactical surprise in a cross-Channel operation of some magnitude' – almost as if this in itself was something of a surprise.[53]

Acceptance of the possible (likely?) loss of tactical as well as strategic surprise might have been defensible in the case of an assault on an open stretch of coast. For Rutter and Jubilee or any other 'frontal' assault without much fire support on a defended position, it clearly would not do. Such an approach betrayed a serious underestimation by the planners of the troops garrisoning Dieppe, partly in terms of strength, for there were more of them in and around Dieppe than anticipated, but also in terms of quality, even though they were thought to belong to a first-class division of the German Field Army (110 ID). The German Army was by a comfortable margin the most professionally skilful army in the field in 1942. With the cancellation of Rutter by C-in-C Portsmouth, over-confidence at COHQ was conflated with a sense of frustration and something resembling a compulsion to launch at least one of the much-touted medium-sized raids. A few small-scale raids had been humiliatingly cancelled by naval Cs-in-C on the grounds that the value of the objective was not worth the risk to the naval covering force. The staff at COHQ were not unmindful, too, that a posting to Combined Operations was considered neither desirable nor fashionable, especially in the Royal Navy. Colonel Head reported to Mountbatten on 8 July that many officers in landing craft flotillas were anxious to return to normal sea duty. Officers in destroyers in the Portsmouth Command deplored raids across the Channel as an unwelcome interference with more pressing responsibilities. The Admiralty had given Baillie-Grohman the impression that their Lordships thought Rutter a dangerous and useless enterprise. As Head put it to his Chief, Combined Operations badly needed a boost.[54]

At a meeting of the Council and Advisers (COHQ) on 7 July, Hughes-Hallett lamented the fact that so far all medium-sized raids had been wiped out at the last minute, usually because of poor weather. Brigadier McNabb thought it possible that exercises like Yukon I were more difficult than the operations they were meant to prepare for. Hughes-Hallett went to the length of suggesting an inquiry by a non-service authority like Sir Stafford Cripps to decide if cancellations of raids had been justified in every case or if the machinery for mounting and launching operations was somehow defective. Perhaps COHQ's meteorological expectations were unrealistic. It was particularly resented by the Advisers that the final authority on the suitability of the weather forecast before an operation was not CCO but the naval C-in-C concerned. The Council resolved that it was 'for consideration' that CCO should be made Supreme Commander for raiding. Mountbatten welcomed this idea but, not surprisingly, 'doubted the advisability' of any inquiry by a non-service authority.[55]

What hope, then, given the fraught atmosphere at COHQ and the undeniable drive and ambition of its Chief, would an Intelligence officer have had of giving pause to the guiding forces behind Jubilee? Realistically, he would have had about as much chance of being taken seriously as Major Unwin, whose warnings about the caves in the cliff-faces of the headlands were ignored.[56] There was no chance of Jubilee's being cancelled except by an Ultra decrypt proving that its security had been compromised or by a decidedly bad weather forecast. Rutter, after all, was not cancelled even after the bombing attack on the two LSIs, but later because of the weather. Nor should it be overlooked, in all fairness, that there was a general consensus in the summer of 1942, shared even by officers who were sceptical about the value of raiding, that a raid on a scale large enough to yield useful lessons was an absolute prerequisite for Roundup in 1943. The weather forecast on 18 August was not especially good, and Hughes-Hallett received a call in the evening from the Admiralty begging him not to sail. But an officer with 'expert local weather knowledge' was most opportunely found, who was prepared to predict a pocket of high pressure moving into the mid-Channel area and lasting until the afternoon of 19 August.[57] That was good enough for the Force Commanders and it actually turned out to be accurate.

There were quite enough reasons already for the cancellation of operations without adding radar to the list. After all, there had been no indication of interception by radar during operations so far, and there was nothing to show that Dieppe was not still in one of the remaining gaps in the Seetakt chain. All the same, COHQ Intelligence warned the planners of Sledgehammer East on 10 July that the enemy was giving high priority to the establishment of an improved ship-searching chain of stations along the coast from Den Helder to Cherbourg. The most authoritative source, R.V. Jones, warned in his Interim Report of 28 July that the list of German stations given in his Appendix was almost certainly incomplete. No Seetakt station, for example, was included for Cherbourg, but it was unlikely that such a sensitive area would have been left unprotected. Twice Jones drew attention to the importance of the coast-watching chain for combined operations. 'In the meantime', he concluded, 'it would be wise to assume that all German occupied harbours are defended by similar stations.' Another document at the end of July pointed out that not all of the coastal stations had been found; there were certainly stations at Blanc Nez and 'probably' near Dieppe and on Guernsey.[58] Wisely or not, the Jubilee planners, who must surely have seen Jones's report, simply ignored the possibility that there was a Seetakt on top of Pointe d'Ailly.

Whether Jubilee *should* have been cancelled because of the threat to tactical surprise posed by radar is essentially a question of operational risk-taking. Deciding what was a 'legitimate' risk is never easy, even with hindsight. Risks that might be acceptable in a small-scale raid become increasingly unacceptable as the scale of the operation grows. In Jubilee's case, the operation was too complex and on too large a scale for the level of support available to justify the risk-taking accepted, especially since strategic surprise was known to have been lost. Radar is merely another, hitherto neglected, reason for saying so. There was a far greater risk of detection than Hughes-Hallett supposed, and diversionary bombing should have been included in any plan for a sizeable landing anywhere on the French coast in the summer of 1942. Had Mountbatten's series of medium-sized raids ever materialized, radar would sooner or later have played a decisive role. In short, the planners of combined operations had enjoyed a run of luck that went back to the unfortunate operator at Chemoulin: that luck held on 18/19 August.

NOTES

1. R.V. Jones, *Most Secret War* (London, 1978), p. 247.
2. Karl Otto Hoffmann, *Geschichte der Luftnachrichtentruppe* (2 Bände, Nekargemünd, 1965, 1968), II, pp. 268–70. KTB, 302 ID, Ia Anlageheft C zum Kriegstagebuch Nr 3: 'Aktennotiz betr Auswertung der St Nazaire Erfahrungen', 24.4.42, T 315 2019. Werner Niehaus, *Die Radarschlacht: Die Geschichte des Hochfrequenzkrieges 1939–1945* (Stuttgart, 1977), p. 93. There was a second Freya at Dieppe by the spring of 1943. Niehaus goes to some length (pp. 108–09) to disprove 'false claims' that Nissen succeeded in removing parts of F 28.
3. Leasor, *Green Beach*, p. 103. Marinegruppenkommando West, Schnellkurzbrief, B Nr gKdos 3867/42 A 1, 28.8.42: 'Geleitsicherung hat Alarmsignal [für Landungsgefahr] nicht geschossen.' T 1022 2363.
4. Leasor, *Green Beach*, pp. 103–4.
5. Nissen, *Winning the Radar War*, p. 165.
6. AIR 16/765, 'Operation Jubilee Questionnaire', P/126A/1 INT, 25 Aug. 1942; CAB 98/22, The Dieppe Raid (Combined Report), Annex 2, Appendix A.
7. See Jones's review of Hinsley, *British Intelligence*, I, *Daily Telegraph*, 27 May 1979; also, letters to *The Times* from Sir Michael Howard and Captain S.W. Roskill, 12 and 19 May 1977 respectively. Hughes-Hallett, 'Memoirs', p. 78: Jones belonged to a 'bunch of very young, very brilliant, and utterly intolerant scientists who attended on behalf of the Air Ministry'. Jones, *Most Secret War*, p. 237, credits Portal for taking him off the raid, not Tizard.

8. See, for example, WO 219/1933, Martian Report No. 10, Annex IV, 'RDF Station at The Hook', 4 Aug. 1942.
9. KTB, 1 Skl, Teil CX, 'Niederschrift über Besichtigungsreise des ObdM nach Frankreich vom 3 bis 6 August 1942', 6.8.42, RM 7/226. The Seetakt was manufactured by Gema and the rate of delivery early in 1942 was only two a month (KTB, 1 Skl, Teil A, 2.4.42, T 1022 1672). The Naval Section (GC&CS) identified 'a station calling itself Ortung-Zentrale' in December 1941 (ADM 223/3, ZIP/ZG/124, 9 Dec. 1941).
10. KTB, BSW: 'Zusammenfassender Überblick', 1–15.8.42, M/444; Frank Reuter, *Funkmess: Die Entwicklung unter der Einsatz des RADAR-Verfahrens in Deutschland bis zum Ende des Zweitens Weltkrieges* (1972), pp. 76–8.
11. KTB, MarGpW: 'Einsatz Stördienst gegen englische Funkmessgeräte', B Nr gKdos 2968/42 A4, 22.6.42, Anlage 1 zum KTB v 1.7–15.7.42, T 1022 3974. KTB, FdS, 21., 22.6.42, T 1022 3253.
12. See, for example, Ruge's twice monthly 'Comprehensive Reviews' in KTB, BSW, M 444; KTB, 2 SichDiv: 'Lagebeurteilung am 16 August 1942', T 1022 3552.
13. The new Lorenz equipments had some initial success but thereafter failed to live up to expectations (KTB, 2 SichDiv: 'Lagebeurteilung am 16 Oktober 1942', T 1022 3553).
14. See the reports from XXV Corps and Naval Group West (both 30 March 1942) included as Anlage 3 and 4 respectively to Oberbefehlshaber West, Heeresgruppenkommando D, Ia Nr 671/42 gKdos, 5.4.42 (KTB C, AOK 15, T 315 502).
15. DEFE 2/126, Operation Chariot, COHQ/Int/II/9 Feb. 1942: this survey of the defences of St Nazaire made no mention of radar, although the subject was raised in a document of 4 Nov. 1941 in this same file.
16. KTB, MarGpW, 24.4.42, T 1022 3974.
17. DEFE 2/65, Operation Bristle; ADM 199/1199, Commodore Commanding Dover, Operation Bristle, No. 580/0215T/42, 12 June 1942; KTB, MarGpW, 4.6.42, T 1022 3974.
18. KTB, MarGpW: 'Einsatz Funkmessdienst bei engl Landung am 4.6.42 in Nähe Le Touquet', FS 2583 A4 gKdos, 7.6.42, T 1022 3974; ADM 199/1199, Operation Bristle, Commodore Commanding Dover, No. 580/0215T/42, 12 June 1942.
19. KTB, BSW, 9/10.7.42, M/444; KTB, 2 Funkmessabteilung Ozet: 'Ortungsergebnisse vom 18–19.8.42, 0800–0800 Uhr', B Nr gKdos

128/42, 20.8.42, T 1022 2429. KTB, MarGpW, 21., 22.7.42, T 1022 3974; KTB, 2 SichDiv, 6., 19.8.42, T 1022 3552.
20. The Naval Staff (Operations) War Diary (KTB, 1 Skl, Teil A, 21.8.42) simply mentioned that the equipment under construction at Pointe d'Ailly was not yet in service on 19 August – seemingly unaware of its temporary installation earlier on.
21. AIR 20/1676, ADI(Sc), 'Air Scientific Intelligence Interim Report: German Coast Watching RDF Stations', 28 July 1942; four copies went to COHQ. AIR 40/1666, Rhubarb Operations: the map at the end of Appendix 10 is dated 15 Jan. 1943; there is no target listed between X/4 (Fécamp) and X/5 (Cayeux).
22. DEFE 2/546, Operation Rutter, 'Minutes of Meeting at COHQ, 15 June 1942', Appendix III. AIR 16/760 contains some useful technical reports on German radar – eg 'German Air Defence Intelligence System', 11G/S 679/Sigs, 27 July 1942; see also AVIA 7/1565, 'Short Report on Enemy RDF in W Europe', AI 4(a), 11 July 1942. GAF equipment was manufactured by Telefunken and the Freya could be mounted on a limber – one reason none had been captured so far in N. Africa. WO 219/1933, Martian Report No. 5, 30 June 1942.
23. KTB, MarGpW, 24.4.42, T 1022 3974.
24. KTB, 302 ID, Ia Anlageheft C zum Kriegstagebuch Nr 3: 'Aktennotiz betr Auswertung der St Nazaire Erfahrungen', 24.4.42, T 315 2019. 'Meldung von Seezeilen durch Funkmessgeräte der Luftwaffe', Luftflottenkommando 3, Höh Näfu Gr Ia/4, Nr 11088/42 geh, 27.4.42, T 1022 2177.
25. KTB, MarGpW, 24.4.42, T 1022 3974; ADM 179/244, Portsmouth Plotting Instructions, 21 June 1943.
26. KTB, BSW, 19.6.42; 31.7./1.8.42, M 444; KTB, 2 SichDiv, 16.7.42, T 1022 3552.
27. Leasor, *Green Beach*, p. 31. KTB 302 ID, Anlageheft D zum Kriegstagebuch Nr 3: 'Gefechtsbericht vom 12/13.6.1942', T 315 2019.
28. AIR 16/760, Baillie-Grohman to Lee-Malory [*sic*], 8 June 1942; Leigh-Mallory to Baillie-Graham [*sic*], 9 June 1942. Baillie-Grohman had arrived from Cairo only on 30 May. Leigh-Mallory sent copies of the letters to Roberts on 9 June.
29. DEFE 2/546, Operation Rutter, 'Minutes of Meeting at COHQ, 5 June 42', Appendix III. AIR 8/895, 'Rutter' Notes for CAS [by Leigh-Mallory]. AIR 16/760, Leigh-Mallory to HQ Bomber Command, 11G/S 500/86, 3 July 1942.

30. AIR 16/746, 'Employment of Bombers by Air Force Commander [for Jubilee]' n.d. ADM 199/1079, 'Naval Force Commander's Narrative', NFJ 0221/42, 30 Aug. 1942, enclosure No. 1.
31. R.V. Jones includes the text of the Oslo Report in *Reflections on Intelligence* (London, 1989), Appendix A, pp. 333–7. Jones, *Most Secret War*, pp. 190–91, 245. AIR 20/1629, ADI(Sc), "DT" (German RDF): Freya and Würzburg', 10 Jan. 1942, Appendix 2.
32. AIR 20/1629, ADI(Sc), "DT" (German RDF): Freya and Würzburg', 10 Jan. 1942; AIR 40/318, ADI(Sc), Memorandum on 'RDF Plotting – Oculist', 25 Oct. 1942, discussed at a meeting at the Air Ministry on 12 Dec. 1942.
33. AIR 20/1629, ADI(Sc), "DT" (German RDF): Freya and Würzburg', 10 Jan. 1942; Jones, *Most Secret War*, p. 122. AIR 16/760, 'German Air Defence Intelligence System', 11G/S 679/Sigs, 27 July 1942.
34. RG 24, C 3, 13,746, War Diary, G Branch, HQ 2 Cdn Inf Div, Oct. 1941–Aug. 1942: 2DS(G)3-3-0, Exercise Yukon II, 22/23 June 1942.
35. Jacques Mordal, *Dieppe*, p. 162. KTB A, AOK 15: 'Grundlegende Bemerkungen des Oberbefehlshabers West Nr 10', Ia Nr 2676/42 geh Kdos, 13.9.42: 'Am Angriffstage zwischen 03.52 und 04.16 Uhr haben Luftwaffengeräte unbekannte Schiffe geortet und sie befehlsgemäss an die Marineortungszentrale durchgegeben.' T 312 502.
36. Hughes-Hallett, 'Memoirs', p. 114. WO 219/1933, Martian Report No. 5, 30 June 1942.
37. Nissen, *Winning the Radar War*, pp. 165, 154–5. Interview with Jack Nissen, 3 Feb. 1988. Nissen's contact was Don Preist, whom he described as an old friend but whose name he misspelled as Priest.
38. DEFE 2/106, Operation Blazing; WO 199/3009. Operation Wetbob, CC(42)42(3rd draft), 30 July 1942.
39. Hoffmann, *Geschichte*, II, pp. 268–70; Leasor, *Green Beach*, pp. 107–8; Mordal, *Dieppe*, p. 161. Some of the confusion might be the result of the transfer of control over the Ozet from 2 Defence Division to 2 Radar Detachment at the end of July.
40. KTB 302 ID, Ia Anlageheft 'Dieppe' zum Kriegstagebuch Nr 3: 'Bericht über den englischen Angriff beiderseits Dieppe am 19.8.42', Abt Ia, Tgb Nr 1105/42 g, 28.8.42, (in particular, 'Zeittafel der Kampfhandlungen'), T 315 2019.
41. Naval authorities, particularly the Harbour Commander in Dieppe,

were critical of the Army for failing to raise the alarm according to Standing Orders ('Notizen anlässlich der Dienstreise des Admirals Frankreich am 24.8.42 nach Dieppe', T 1022 2363).
42. The War Diary for 302 ID for 1942 contains several examples of the steps taken in response to a warning of a surprise attack based on an agent's report, intercepted telephone conversation, etc; Ia Anlageheft D contains several 'Gefechtsberichte' – on 3/4 April, for example, 'Gefechtsbereitschaft' was ordered in advance for IR 571, FMZ etc, while the divisions on either flank were sent an 'Orientierung'. T 315 2019. Atkin, *Dieppe*, pp. 115–16, 133.
43. Mordal, *Dieppe*, p. 160; Hoffmann, *Geschichte*, II, pp. 268–70; Leasor, *Green Beach*, pp. 107–8.
44. KTB, BSW, 25/26.7.42, M 444; KTB, 2 Funkmessabteilung Ozet: 'Ortungsergebnisse vom 18–19.8.42, 0800–0800 Uhr', B Nr Gkdos 128/42, 20.8.42, T 1022 2429.
45. KTB, 2 SichDiv: 'Seegefecht vor Dieppe 11/12.12.42.', Führer der 2 Sicherungsdivision, B Nr Gkdos 3313, A3, 29.12.42, T 1022 3553.
46. AIR 40/2239, 'Report by A W Bonsall', 30 Aug. 1942, on Y Intelligence for Combined Operations. *Daily Telegraph*, 7 Oct. 1991. Leasor, *Green Beach*, pp. 154–5. According to Jones (*Most Secret War*, pp. 195–6) much of the early credit for Oculist belonged to the Army intercept service.
47. DEFE 2/324, Mountbatten to Crerar, Letter 10546, 5 Oct. 1942; Crerar to Mountbatten, 13 Oct. 1942.
48. DEFE 2/324, Note with indecipherable signature, 1 Oct. 1942. Hughes-Hallett rang CCO on 1 October to assure him that the convoy coming down the coast had entered Dieppe at 0330; and that the craft encountered by Group 5 were never plotted. DEFE 2/324, J.D. Cockroft to J.D. Bernal (COHQ), 11 Nov. 1942.
49. To keep the Germans guessing about the effectiveness of their jamming, K stations were never operated without some CD/CHL stations providing cover, although CD/CHL stations were at times operated on their own; thus the Germans could not be sure whether there was an alternative to the CD/CHL or whether an anti-jamming device had been devised for it (WO 199/534, HF/6456/18/RA(CA), 2 April 1942).
50. DEFE 2 324 contains the reports to Cockroft from his staff at ADRDE which he forwarded to Bernal with a covering note on 11 Nov. 1942. ADM 234/355, Raid on Dieppe (Naval Operations), 19 August 1942, Naval Staff History BR 1736 (26), Battle Summary

No. 33, (1959). Villa, *Unauthorized Action*, p. 16, implies that Hughes-Hallett turned a Nelsonian blind eye – 'It has been argued that there was a failure in signal transmission [to *Calpe*] . . .' but the real failure was on the part of *Brocklesby* and *Slazak* to pick up the signal and disperse the German ships.

51. According to 'RDF Policy for Jubilee', radar silence was to be observed on Naval Types 286 and 290, while Type 272 was used as a protection against surface attacks; operators of Types 286 and 290 were to disconnect transmitting aerials and set a listening watch for Type 78 beacon signals from the two MLs. See ADM 199/1079, 'Naval Force Commander's Narrative', NFJ 0221/92, 30 Aug. 1942, enclosure No. 1. Hughes-Hallett ('Memoirs', pp. 169–70) found that GEE was extremely accurate in mid-Channel but became progressively less accurate east or west of there 'and would not have done for the particular purpose for which we required it'.
52. WO 106/4142, 'Security of Combined Operations Base: Note by Lord Swinton', 20 May 1942, including reference to a letter from Mountbatten of 7 May 1942.
53. AIR 16/760, Leigh-Mallory to Baillie-Graham [*sic*], 9 June 1942.
54. DEFE 2/549, 'Draft Notes on the Preparation and Mounting of Operation Rutter', by Military Adviser to CO, 8 July 1942'; Baillie-Grohman, 'Planning Notes on Rutter', GRO/28, Baillie-Grohman Papers.
55. DEFE 2/564, Minutes of Meeting of COHQ Council and Advisers, CCO in Chair, 7 July 1942.
56. Terence Robertson, *Dieppe*, pp. 217–18.
57. See, for example, 'Raids on the French Occupied Coast', Memorandum by the Naval C-in-C Expeditionary Force, CC(42)47, 24 July 1942 (copy in DEFE 2/306), in which Admiral Ramsay criticized the policy of raiding: 'The Germans no doubt welcome these raids, for nothing shows up weakness in the defence more than an attack with a very limited objective.' Hughes-Hallett, 'Memoirs', p. 177.
58. DEFE 2/561, Sledgehammer (East) Part I. AIR 20/1676, ADI(Sc), 'Air Scientific Intelligence Interim Report: German Coast Watching RDF Stations', 28 July 1942.

5

Most Secret and Other Sources

When in 1974 Group Captain Winterbotham broke the oath of lifelong secrecy he himself had administered to many of the wartime recruits for the GC&CS at Bletchley Park, he caught professional historians and the general public alike by surprise. The Ultra secret had been exceedingly well kept. There was no hint of Ultra intelligence as such, or under the rubric of Special Intelligence (SI), Most Secret Source (MSS), or Boniface, in any of the volumes in the Grand Strategy and Campaigns series of the British Official History. Ultra, as Sir David Hunt put it, was 'the missing factor in every description of strategy to those [like himself] in the know'. The trouble was that Winterbotham had merely let the cat out of the bag; his book, necessarily written from memory, was inevitably inaccurate in places and hardly a fitting tribute to the great British success story of the Second World War. Accordingly, when the first batch of 25,000 decrypts and other Ultra-related material were abruptly released at the PRO in October 1977, the very least that was expected was a wave of revisionist history at its best.[1]

There has indeed been an outpouring of history, some of it at its best, since then, thanks largely to Bletchley Park veterans and others who kept the secret during the war and for thirty years thereafter. The first researcher through the doors of the new PRO building at Kew in 1977 was the Cambridge historian Ralph Bennett, who had served as an Army duty officer in Hut 3; his subsequent publications include *Ultra in the West* (1979) and *Ultra and Mediterranean Strategy* (1989), based on some of the material he had helped to evaluate. Others of the same semi-official standing were Patrick Beesly and Peter Calvocoressi, formerly of the Operational Intelligence Centre (OIC) at the Admiralty and the RAF section in Hut 3 respectively. Hinsley joined GC&CS as a 'third-year man' in History from St John's College, Cambridge, in 1939. Thanks in good part to their efforts, Ultra's value for the Allied cause has been put into sharper focus than was possible in 1974. Historians have generally arrived at a rough consensus that Ultra undoubtedly shortened the war against Germany and Italy and had a profound impact on several of its decisive campaigns, but certainly did not win it. In Bennett's opinion, Ultra had its greatest effect on military and air operations at a level 'slightly below that of grand strategy, although it [Ultra] made periodical highly successful excursions into battlefield tactics'. Significantly enough, though, he was doubtful in 1981 that even the release of the entire corpus of decrypts would 'compel so drastic a revision' as was so widely forecast with the publication of Winterbotham's bombshell.[2]

Whether those former members of the wartime intelligence community who turned to history after 1974 should be categorized as 'revisionists' is open to question. Their publications reflect their frequently repeated observation that the work of intelligence, especially the evaluation of Sigint in unprecedented volume, lay mainly in the patient sifting of detail about the enemy. It was extremely rare, according to Calvocoressi, to be presented with a plum on a plate: either it was not a plum or it was not on a plate.[3] The revisionist label is better attached to outsiders like Anthony Cave Brown, who was very quick off the mark with his book in 1975, and whose claim that the exploitation of Ultra was at times inhibited by excessive regard for its security has proved to be quite resilient. Phillip Knightley repeated it in 1986, though without insinuations about conspiracy. James Rusbridger, a leading exponent of the idea that not enough attention has been paid to the successes of German Intelligence, finds it hard to accept the official line that German cipher security experts failed to grasp that at least some of their Enigma keys had been broken. So far, however, neither he nor Knightley has produced much in the way of documentary support.[4]

The question of Ultra's involvement with Jubilee has not attracted much attention at all. Hinsley's Dieppe Appendix spent more time on PR and low-grade Sigint generated during the operation than on Sigint of the highest grade. It would appear from elsewhere in his text that Ultra played a greater role at St Nazaire than at Dieppe. The OIC 'determined the approach route and the timing of the raid [Chariot] in the light of the Enigma intelligence about the enemy's swept channels and recognition signals . . .' According to Patrick Beesly, SI bought the raiding force a few vital moments before surprise was lost. 'OIC was less successful', he adds (not altogether accurately), 'at the time of the Dieppe operation, when an encounter with an unexpected convoy alerted the German defenders before H hour.'[5]

Yet it would be a mistake to conclude that this is a subject devoid of interest or out of reach for lack of source material. The German Navy's Heimisch key was producing hundreds of decrypts a day in 1942, all of which have been available since 1977. Ultra's direct involvement with Rutter/Jubilee, for example, began with the following message from the OIC to Cs-in-C Portsmouth, Plymouth and The Nore and Vice-Admiral Dover on 27 June: 'German air reconnaissance of the south coast up to 22 June has established a great increase in the number of small craft suitable for landing operations. The main concentrations are between Selsey Bill and Portland, and between Brixham and Plymouth.'[6] Other relevant keys were doubtless productive but their yield of decrypts remains as yet unavailable. Fortunately, however, there are just enough supplementary documents, mostly originating in the OIC, with references to decrypts belonging to unreleased series, such as the GAF Red key, to warrant tackling the question of Ultra and Jubilee. Of primary interest, of course, was how far Ultra could have been relied upon to provide warning that the enemy had discovered the time and place of this or any other impending cross-Channel operation.

A key document was issued by the Naval Section at Bletchley Park on 15 August: 'Increase of German Defence Measures, Western Area 1942' (ZIP/ZG/173). As might be expected, this survey included a great deal of detail about defensive minelaying, the closing of ports by booms or cables, off-shore patrols, the ground defence of aerodromes, and so on – measures often introduced on an emergency basis after St Nazaire. The handful of recipients on the distribution list were reminded that the account was 'incomplete but symptomatic' since it was 'dependent for most of its evidence on accidents and circumstances which decided the issuing of [W/T] signals'. What it did provide was proof of 'the steady and already considerable progress' the Germans had made to counter a landing in the West, including Norway, Denmark and Holland. More to the point, it also provides, when compared with German sources, a ready check on the occasions when the German defensive machine reacted to the threat of an Allied operation.

There was every indication, according to the Naval Section, that 'from time to time' the Germans took the threat of invasion quite seriously, although Allied propaganda and rumour rather than visible preparations prompted German steps to increase preparedness. The first alarm took place in the north of Norway at the end of April, according to decrypted signals sent by Fliegerführer Lofoten. A fully-fledged invasion scare flared up in Norway in the middle of May, when an agent's report resulted in a state of 'Alarm Readiness' for all three services in the Norwegian area. Clearly, Operation Hardboiled must have had a part to play in this, although there was no mention of it. This particular flap was not confined to Norway, because on 18 May the following message was addressed to the Commander of Alderney:

> During the present week, the night hours and the phases of the moon are favourable for attempts at landing by the enemy. Alarm exercises are therefore to be carried out at this period, the times of which are to be varied in such a way that the troops are ready an hour before moonrise. With the beginning of the next moon period the danger will be past. The troops must (as a result of the exercise) be ready to operate immediately in the event of a landing. Therefore only live ammunition is to be carried [CX/MSS/1025/T4].

At the end of May, orders from the Operations officer of KG 30 for the operational use of the cadre of the divebombing training unit at Aalborg underlined the state of alertness against a landing on the Danish coast. There were two 'principal alarms' in June. Special watchfulness was ordered in the Channel area on 12 June 'as according to a Secret Service [Abwehr] report, the English intend to make a landing on the French coast on the night of

12-13/6 [CX/MSS/1087/T12]'. On 16 July, naval authorities in Granville announced a state of imminent danger 'for tonight and subsequent nights with immediate effect [STPG 4090]'. Two days later FOIC Channel Coast (Marinebefehlshaber Kanalküste) warned the Channel Islands that 'several agents' reports received today at the Rouen reporting centre (of Army Group D) are to the effect that the rumour is circulating in pro-British circles in Rouen that the British will undertake large-scale landing attempts from 0400/18/7 along a wide stretch of coast. Rouen reporting centre submits the reports with all reserve [ZTPG 65952] . . .' Finally, as proof of the state of alertness early in August, Fliegerkorps XII issued the following on 9 August:

> According to the opinion of the highest quarters, the political situation on the enemy side has become so acute owing to the great German successes in the East that the probability of an Anglo-American landing in the immediate future has become very great. For this reason the greatest alertness and readiness is indispensable on the part of all subordinate authorities and units. Attention is emphatically drawn once again to the orders published with regard to defence security and anti-sabotage measures [CX/MSS/1283/T18].[7]

The Naval Section's document explains a great deal. The 'principal' anti-invasion alarms and exercises were indeed registered by Ultra. The Abwehr reports of 4 and 13 August set off no such reaction; and, interestingly enough, there was no mention either of Hitler's Directive No 40 or Order of 9 July. Obviously, whether Ultra could provide warning in time for the cancellation of an operation whose security it revealed to have been compromised, depended entirely on the speed with which the signal could be decrypted. There was only a matter of hours between the original alerts on 12 June and 16 and 18 July and the anticipated landing on the coast. But provided a warning was relayed in time to COHQ, CCO had the authority to cancel an operation or recall it before the point of no return in the Channel was passed – 'on account of special information that might have become available'.

At a meeting of the COHQ Examination Committee on 9 June, for example, Casa Maury, Senior Intelligence Officer at COHQ, stated that a report from a first-class source revealed that the Germans had received information of a possible raid on Blazing (Alderney) and were on the alert for it; he agreed for a memo to this effect to be circulated to all holders of Blazing cards. Casa Maury's first-class source was very likely CX/MSS/1205/T4 of 18 May, though it is extremely unlikely that the original German signal to Alderney was the result of leakage of information about the raid cancelled earlier in May. Blazing was at one stage planned to employ a parachute battalion as well as a seaborne force and the wording of the signal certainly suggests that the Germans were worried about an airborne operation. On the other hand, they had generally become very parachute-conscious since Bruneval. Rather, this decrypt was probably an example of one of the 'accidents and circumstances' upon which the use of W/T depended.[8] The invasion scare in Norway spilled over into the entire 'Western Area' in the second half of May. Therefore this signal to an island garrison whose other lines of communication to the mainland were subject to interruption may have had less specific significance than might at first have appeared. Almost exactly a year later, the same sort of thing happened again. Following a 'Dear Dickie' letter from 'C' on 3 May 1943, Operation Coughdrop, a raid to destroy U-boats inside their pens at Lorient, was scrapped. 'A German Secret Service report', wrote Menzies, 'indicates that operations directed against the submarine base at Lorient are expected by the local inhabitants.'[9] His source was almost certainly the Abwehr Enigma.

In the light of what happened with Blazing and Coughdrop, it is interesting, to say the least, to come across a confident claim – 'hard and fast evidence', allegedly – that Ultra had provided proof of German foreknowledge of Jubilee 'as early as 12 August'. This appeared in a USAAF report by a Lieutenant Colonel Haines, which was compiled in September 1954 but not published until 1980, and dealt with the contribution Ultra made to the US strategic bombing offensive. The repetition of the phrase – 'hard and fast evidence' – and exact coincidence of the date strongly suggest that this publication, if not the same as the '1945 monograph' referred to by David Irving in the *Daily Telegraph* in 1989, at least shared the same source. Ziegler's *Mountbatten* (1985) also touched upon a possible connection between 'Ultra intercepts' and 'some foreknowledge' of the raid on the part of 'elements of the German high command', citing NSA and a history of the US strategic air force. Regrettably, Haines provides no references and the text of the decrypt in question is not given. Still, the claim should not be rejected out of hand, if only because it has something of the authority of a primary source, Haines apparently having enjoyed access to a selection of Ultra decrypts and other material that has so far not been released. He used the Air Ministry's 'chronological master file' of all Ultra messages in the sequence in which they were received by AI and 'diplomatic' decrypts in the ISOS (Abwehr) series. He also had access to SR reports of monitored POW conversations and other AI MSS files arranged according to subject matter.[10] Assuming for the moment that his claim is valid, it seems strange that the evidence on which it was based was evidently left out of the Naval Section's paper.

Luckily, not all references to Ultra or MSS were weeded from the documents before their release in 1972; some AI files contained a number of 'Destroy by Fire after Reading' documents or impressively detailed items of intelligence that must have puzzled researchers in the two years before *The Ultra Secret* appeared. On 6 June 1942, for example, AI 3(e) sent a memo to the Assistant Chief of the Air Staff (Intelligence) based on MSS. A German police decrypt had disclosed that Hitler had a HQ in or near Dnepropetrovsk by referring to the transfer there on 26 May of replacement personnel for 'ARP Abteilung (mot) 16' currently at the 'Führerhauptquartier'. Hitler, in fact, visited Poltava on 1 June and later set up an HQ at Vinnitsa.[11] More often, the MSS source was 'sanitized' by the use of expressions like 'a first-class source' or 'special information that has become available'. At any rate, there survived in AIR 16/764 a copy of an undated, Most Secret note from Casa Maury to the three Jubilee Force Commanders:

> Information which is considered reliable has been received reporting that German authorities regard an Allied landing in the West, provoked by German successes in the East, as likely to be attempted at any moment. All defences to be armed and held in instant readiness. The area involved is Holland, Belgium and northern France.[12]

Its placement in the file suggests that this note was received at HQ 11 Group between 11 and 13 August; and its wording contains a very distinctive resonance. Hitler's Befehl of 9 July began with a flourish about German successes in Russia: 'Die schnellen und grossen Erfolge im Osten können England vor die Alternative stellen . . .' Thereafter, this particular wording to justify various anti-invasion precautions became something of a refrain, not least because German successes in Russia continued through July. The 302 ID War Diary entry listing the telephone relay from LXXXI Corps at 0235 on 10 July began: 'Auf Grund der grösseren Erfolge im Osten und anderen Anzeichen, rechnet OB West mit feindl. Landungen . . .' On 17 July, the staff of 302 ID used roughly the same wording as a preamble to Divisionsbefehl Nr 96 für den Küstenschutz.[13]

Casa Maury's note was actually a sanitized derivative of CX/MSS/1283/T18 of 9 August. The timing is right but what gives the game away are the matching references to 'German successes in the East'. The originating authority was Fliegerkorps XII, whose operational area happened to coincide with 'Holland, Belgium and northern France', and the signal was a circular to subordinate units on the eve of the next period of 'drohende Gefahr'. Fliegerkorps XII's Enigma key (Cockroach) had been broken on the first day of use, in January 1942, and was read until the end of the war. There was no real reason to call Jubilee off because of this intelligence, which mentioned neither Dieppe nor 19 August. The possibility that the USAAF study was referring to a different decrypt altogether must be considered remote, since there was time for it to have been included in the Naval Section's paper of 15 August. Perhaps the last word on the subject of Ultra and alleged German foreknowledge of Jubilee is better left to Commander Ian Fleming RNVR of the Naval Intelligence Division (NID) at the Admiralty than to Lieutenant Colonel Haines. Reporting to the Director of Naval Intelligence on 21 August about his experiences on board *Fernie* at Dieppe, Fleming listed the usual reasons why there had been no security leakage. His report is to be found in one of the COHQ files opened in 1972; therefore his last reason was another of those pre-Winterbotham references to Ultra: 'No indications from Most Secret Sources point to surprise not having been effected.'[14]

Apart from providing insurance against a breach of security,

Ultra had the potential to influence operations in general in two ways: by providing intelligence fast enough to have an immediate tactical application; and by offering useful background information about the enemy's order of battle, dispositions, standard operating procedures, and 'patterns of behaviour', as Beesly put it. The brief duration of Jubilee and the sheer volume of other forms of Sigint seem to have ruled out any direct bearing the GAF Enigma might have had on the air battle. But the slower pace of naval operations, not to forget the availability of the decrypts, makes SI a more promising field of inquiry. SI was an important factor in naval operations in the Channel, thanks to GC&CS's reading of Heimisch, the key used by ships and shore authorities in home waters, with 'virtual currency' since August 1941. Much of the traffic was repetitive and administrative in content, but with the help of other W/T intercepts and PR, it was possible to determine what was normal activity between the Elbe and the Gironde, and thus what was out of the ordinary. The Admiralty, it should be remembered, was in a position to exercise direct control over operations in Home waters, and the OIC, although it had no executive power, had the authority to communicate directly with the Fleet at sea and naval Commanders-in-Chief.[15]

The first signal in this key triggered by Jubilee was a Most Immediate from the Naval Communications officer in Le Havre: 'Attack on convoy at 0450 4 miles off Dieppe by surface forces [ZIP/ZTPG/68913].' The Time of Origin (TOO) of the signal was 0510 German time (0410 BST), and the time of arrival of the decrypt at the OIC was 0710 on 19 August. Three hours to cover interception, decryption, translation and transmission by teleprinter to London was very good indeed, given that delays of a few days were not uncommon in decrypting signals in the Heimisch key. The Naval Signal Station in Dieppe had passed this information by landline to Le Havre (and also to the FMZ and Hafenkommandant in Dieppe) and it was being relayed by W/T from there to Naval Group West, 3 Air Fleet and Army Group D when intercepted. The first Enigma signal about an actual landing originated with the Sea Defence Commandant Seine-Somme and announced that the enemy had landed at 'Bernevale [sic]' at 0525 and that fire had been opened on enemy ships. This signal (ZIP/ZTPG/68930) had a TOO of 0704 (0604 BST) and

DIEPPE REVISITED: A DOCUMENTARY INVESTIGATION

PLOT OF ENEMY CONVOY, 18-19th. AUGUST 1942   PLAN 5

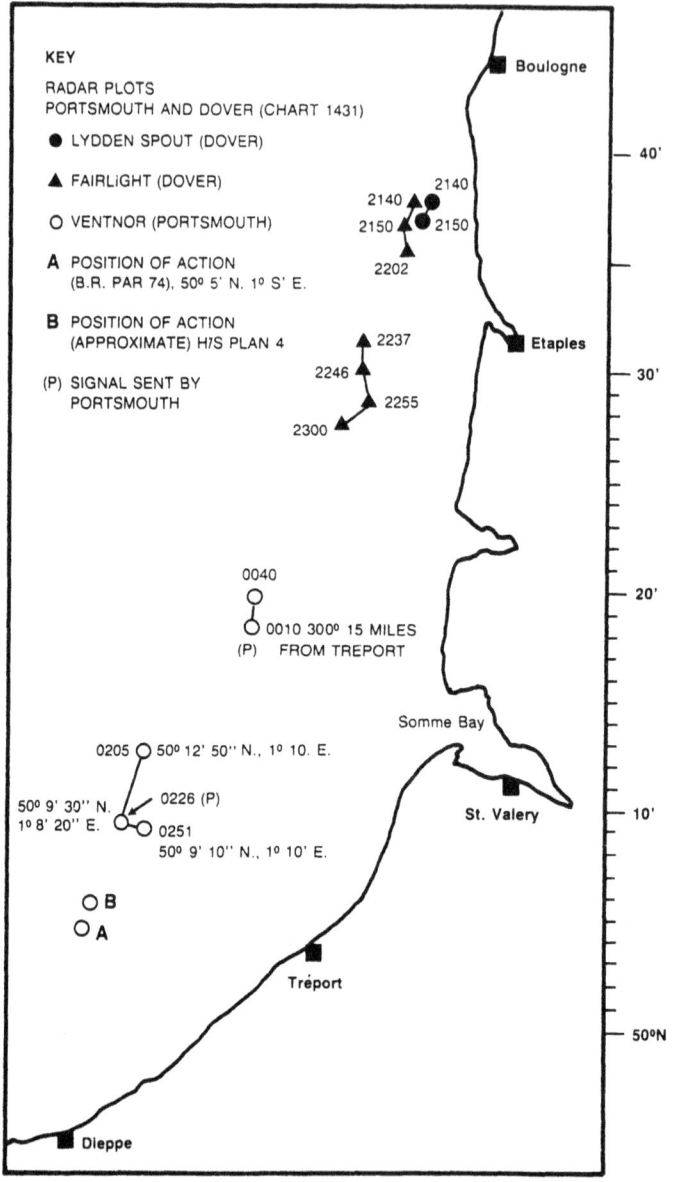

Map 3   Plot of Enemy Convoy (Crown copyright)

reached the OIC at 0726 GMT (0826 BST). Later that morning SI provided plenty of information about the fate of Convoy 2437. By 0641 UJ 1411's wireless had been repaired, enabling Wurmbach to transmit the following signal: 'Convoy dispersed north of Dieppe owing to enemy landing. UJ 1404 hit in magazine has vanished burning fiercely. One gunboat destroyed by ramming. Several hits on MTBs and on 1 Flotilla leader. 2 a/c shot down. Convoy gathering at Le Treport. UJ 1404 and 1 motor vessel are missing. W/T watch on coastal M/F again [ZIP/ZTPG/68938].' Less than a quarter of an hour later, UJ 1411 reported that the convoy was anchored in Le Treport roads while she and M 4014 were heading back to Boulogne (ZIP/ZTPG/68941). Those two decrypts reached the OIC at 0949 and 1003 respectively. At 0823 2 Defence Division sent a final, despairing signal to UJ 1404: 'Immediate. Proceed Le Havre [ZIP/ZTPG/68956].'[16] The point is, of course, that fast as the times achieved by GC&CS were, by the time the decrypts reached the OIC they no longer had any current tactical value.

The Germans routinely signalled in advance the time of sailing, composition and destination of their coastal convoys. The two convoys in either direction between Boulogne and Dieppe on 18/19 August both left at 2000 and were expected to arrive at 0500, travelling at six knots. The commander of 15 Patrol Boat Flotilla signalled his convoy at 1839 on 18 August and the decrypt reached the OIC half an hour before his scheduled time of arrival in Boulogne. Convoy 2412 was well clear of the operational area by midnight. The TOO of UJ 1411's signal from Boulogne was 1952, 18 August: 'Leaving Boulogne 2100 for Dieppe with UJ 1404, M 4014 and convoy of 5 (corrupt group ? + motor)ships. Speed 6 knots. Coastal H/F West.' This decrypt (ZIP/ZTPG/68850) reached the OIC at 0316, or just half an hour before the collision with 5 Group.[17] No signals were sent by C-in-C Portsmouth to the Naval Force Commander after 0244, when the second warning of enemy units plotted by shore-based radar was transmitted on the Admiral's waveband. Is this one of those alleged instances when the full use of Ultra was inhibited by excessive regard for its security?

At the most general level, it has perhaps become too easy to forget how extraordinarily tight the security surrounding Ultra was during the war. The First Lord of the Admiralty and many officers within NID were not 'indoctrinated' to receive Ultra. There could be no question of the casual use of SI to answer routine questions – which presumably explains, for example, why the COHQ investigation into the radar events of 18/19 August persisted in the story of a 'tanker and escort'. Mountbatten himself was, of course, indoctrinated and as a member of the COS received a daily headline sheet of choice Ultra intelligence circulated by 'C'. But very few officers in his Headquarters and none of the officers who found themselves on their way to raid Dieppe – except apparently for Fleming, the DNI's personal representative, who was not allowed to go ashore – were on the Ultra list. The First Sea Lord had directed that no operation based on SI should be launched unless this could be done without endangering the source. German suspicions could easily be aroused by action inspired by SI in circumstances where there were no other plausible sources of intelligence. For example, no major units of the German Fleet whose departure from the Baltic had been predicted by SI should be attacked unless they had first been reported by aerial reconnaissance, agents or submarine. When Ultra signals were transmitted to units at sea, there had to be an officer of sufficient seniority (Flag rank) on the receiving end who was also indoctrinated. Only in very exceptional circumstances did the OIC send a signal based on SI directly to unindoctrinated recipients at sea, and then only after carefully paraphrasing the contents and giving serious consideration to discouraging speculation about its source by the recipients. One of those rare occasions took place within days of Jubilee when undiluted SI led directly to the sinking of the minelayer *Ulm* off Bear Island.[18] The whole point about Ultra was that it was not a fragment of Sigint which was unlikely to be repeated and therefore exploitable, but a vast volume of reliable, timely and continuous information which should not be risked for short-term operational gain.

The use of Ultra in communications, say, with the Home Fleet in support of the convoys to north Russia, is easy enough to understand. Operations in the Channel, however, were a different proposition. Beesly made it clear that secure channels of communication had been set up for the distribution of SI by the OIC to Senior Intelligence Officers, SO(I)s, at the naval Home Commands, but he added that few records of this aspect of the OIC's work have survived. Speed was guaranteed by the use of the green scrambler telephone, presumably followed up by an 'Immediate Ultra' on the teleprinter. According to Hinsley, SI was seldom provided with sufficient currency to make it possible to intercept raiders or blockade breakers on passage by stages up- or down-Channel. When Raider B (*Komet*) was sunk off Cherbourg by destroyers on 13/14 October 1942, the relevant Enigma traffic was subject to a 36-hour delay; it was the experience of reading months of German signals that enabled the OIC to predict the route Raider B would follow. Then an 'invaluable' aerial reconnaissance report made it possible for C-in-C Plymouth to warn the destroyers at 2359 on 13 October that the enemy would be close to Cap de la Hague at 0045/14.[19] SI could not be relied upon to guard against a chance encounter during a raid. At Bruneval, the landing craft waiting to pick up the parachutists sighted enemy destroyers or torpedo boats moving on a westerly course along the coast. Had the Germans appeared an hour earlier, the whole operation could have been ruined. Yet the OIC had assured C-in-C Plymouth by telephone on 27 February that there were no patrols in the vicinity. In fact, the landing craft had spotted a minesweeper, two R-boats and an R-boat depot ship on their way from Boulogne to Le Havre, but this was not revealed by SI until 1 March.[20]

The precautions observed in cases where SI gave advance notice of shipping movements in the Channel fast enough to have operational application are difficult to reconstruct. But it is safe to assume that the handling of SI by SO(I)s was subject to no less stringent terms than those enforced by the OIC itself. In any case, there are overwhelming circumstantial reasons why no additional warning was transmitted after 0316 to alert Hughes-Hallett to the possible proximity to Dieppe of Convoy 2347. In the first place, two signals had already been sent to *Calpe* pointing out the presence of enemy vessels to the eastward, and there was no reason to suppose that she had not received them. Secondly, Convoy 2347 should have been easy prey for *Slazak* and *Brocklesby*. Thirdly, any such W/T signal would have been based exclusively on SI and carried the risk of interception and decryption by the B-Dienst. The evidence is lacking to answer the interesting but hypothetical question: What if SI had disclosed the possibility of a collision with

E-boats? At 2313 on 29 July, for example, Cs-in-C Plymouth and Portsmouth were informed by an 'Immediate AIDAC' that '3 E boats will be proceeding Cherbourg to Guernsey between 0400 and 0500 B/30'. Therefore, although according to Hinsley there was an average delay of 15 hours in producing the decrypts giving details of proposed E-boat sorties, there were occasions when advance notice of their activities was received fast enough to be of operational value. On 18/19 August, the timing was fast but not fast enough: at 1146 on 19 August, the following decrypt reached the OIC:

> From: Sea Defence Commandant Channel Islands
> To: Naval Harbourmaster Alderney
>
> (4?) S boats will leave Cherbourg westbound at 2200, returning through the Channel Islands. They will put into Cherbourg about 0530, but if weather conditions are unfavourable they will put into Guernsey [ZIP/ZTPG/68969].

The four boats in Boulogne were not heard from, on the Enigma at least, until pm on 21 August, when it was revealed that they had been ordered to comb the Channel from Etaples to Fécamp at 31 knots on 19/20 August. SI also showed that three of the Cherbourg boats had covered the Bay of the Seine on a similar mission, putting into Le Havre at 0600 on 20 August with a Spitfire pilot they had picked up. SI was particularly valuable in keeping track of E-boats, since it was not always possible to distinguish between them and R-boats in PR photos, and pens had been constructed to shelter them at Cherbourg and Boulogne.[21]

So far as Ultra's second possible contribution to the success of an operation goes, namely, the provision of in-depth intelligence about the enemy's order of battle and latest dispositions, the story is more straightforward. At 1620 on 18 August, just before sailing time, C-in-C Portsmouth received from the OIC a comprehensive survey based on SI of German naval dispositions between Ostend and St Malo as at noon on 17 August. There were the eight E-boats, four each at Cherbourg and Boulogne and seven 'M/S trawlers' (Patrol Boats) at Dieppe. UJ 1404 and UJ 1411 appeared as two 'A/S' (anti-submarine craft) at Le Havre. Among the probable movements forecast for 17/18 August was a convoy from Le Havre to Boulogne, though the two A/S were not included as escorts; as German sources were to show, it was only by a last-minute arrangement with 14 UJ Flotilla that they sailed that night for Boulogne. Another probable convoy was from Dieppe to Boulogne – a 'tug- and lighter-convoy escorted by 7 M/S trawlers' – which also sailed on schedule. SI had kept track of the considerable movement of E-boats since 13 August, when the redeployment from the Channel to the North Sea began. Seven boats left Cherbourg on 14 August for Ymuiden, with two being detached to Boulogne. As late as 15 August there were as many as nine at Boulogne, but their movements on 15/16 August left only four each at Boulogne and Cherbourg, as duly disclosed in an OIC update on 17 August. A PR survey of all the Channel ports was flown on the afternoon of 18 August to bring intelligence about naval dispositions completely up-to-date.[22]

The promulgation of enemy naval dispositions on the eve of an operation seems to have been standard OIC procedure. Although neither Beesly nor Hinsley said so, there can be little doubt that the OIC also played a role in the planning of Jubilee, as the main source of intelligence about German minelaying. SI analysis of German swept channels in the North Sea had led to the mining of *Scharnhorst* and *Gneisenau* in the final lap of the Channel Dash in February; in March, the same source had an important bearing on the planning of Chariot. According to Beesly, decrypts indicating a change in the normal patterns of German behaviour had a way of ringing alarm bells in the minds of 'Denning's experts'.[23] The documents show that on 10 July, one of those experts predicted, by a neat piece of deduction, the laying of the flanking minefields in mid-Channel. According to the decrypt in question, the Senior Officer 2 Defence Division had arrived at Lorient by car late on 8 July to take command of 8 M/S Flotilla; a minesweeper had been sent to Lorient to fetch him and his writer. It was known that 8 M/S and 2 M/S Flotillas had been carrying out minelaying exercises in the Quiberon area, and since the area of command of 2 Defence Division was from Flushing to Cherbourg, the OIC inferred that there was a likelihood of minelaying in the Channel 'in the near future'. On 11 July the OIC advised Cs-in-C Plymouth, Portsmouth and The Nore and VA Dover: 'At least 10 M Class Minesweepers have recently been carrying out minelaying exercises by night in the Bay of Biscay. There is reason to believe that they intend to operate in the Channel area.'[24] In an impressive demonstration of the co-ordination of various sources of intelligence, first their arrival at Cherbourg was reported by PR, then the first sortie by six M Class minesweepers on 11/12 July was plotted by radar, and finally the laying of the suspected minefield was confirmed on 15 July by the decrypt of a signal specifying the co-ordinates of the area north-east of Cherbourg now to be considered dangerous even for shallow-draught vessels. The Naval Section's paper on German Defence Measures included an Appendix on minelaying. It could scarcely have been coincidental, therefore, that the two British minesweeping flotillas swept no mines as they cleared the two channels through the barrier for the LSIs, destroyers and other vessels. So much for the rumour current in Whitehall early in August that the Channel had been heavily mined and serious losses were to be expected in any major combined operation from then on.[25]

The captured portions of Jubilee Naval Operation Orders (NFJ 0221/92, 10 August) handed German Naval Intelligence unsettling evidence of how well informed the enemy was about certain features of their coastal defences. The landing force had approached the coast through one of the last remaining gaps in that part of the minebarrier. The Operations staff at OKM commented on the quality of the navigational sketches, which they found almost good enough to suggest that the British had been party to the planning of the 'Sperrsystem'. Group West admitted that there was a striking correspondence between the 'suspected' minefields in the captured charts and the 'Sperrsystem' as a whole.[26] Why, then, did they not raise the question of cipher security? Quite simply, there were too many other possible sources of intelligence, given the operational conditions in the Channel. In the case of the gap in the minebarrier (mistakenly identified at first as between E 2 and F 5), there was a ready explanation at hand in the superiority of British radar, whose accuracy and resistance to jamming had already rendered offensive minelaying both pointless and hazardous for E-boats. German naval staffs were also aware that their defensive minelaying in mid-Channel or closer to the French coast was within range of stations on the south coast. The B-Dienst station at Brughes (MP Abteilung Flandern) routinely monitored MTB R/T on

VHF as well as coded W/T signals from Dover, Portsmouth and Devonport reporting plots of German movements to MGBs and MTBs at sea, and the precise movements of minelayers made them an easy plot.[27] The Germans, however, found consolation in the idea that the minebarrier was intended to have a deterrent effect anyway; it was also slightly encouraging that not every minelay was reported on the MTB waveband. The T-boats that came within 21 miles of the English coast on 20/21 July were apparently missed altogether; rather more surprisingly, 26 R-boats operating out of Boulogne from 0345 to 0615 on 15 August went unreported by Dover.[28]

Of course it was a matter of guesswork whether these operations had been detected but not reported or not detected at all. In the OIC SI Summary covering Channel activities from 11–15 August (OIC/SI 312, 19 8 1942), the R-boat sortie under the command of SO 2 Defence Division – 'this officer commanded the M Class minesweepers when they were engaged in minelaying recently' – was itemized, including the precise co-ordinates of the mined area that was declared dangerous later that day even to shallow-draught vessels. Interestingly enough, while the minelays carried out by 8 M/S Flotilla from Le Havre on 19/20 and 20/21 July were recorded by SI, the participation of the two T-boats was not, although it was known from SI that four boats of 3 T-Flotilla had arrived at Le Havre on 16 July. The decrypted German Naval Summary for the Channel and Atlantic areas for 21 July reported that the two minelaying operations had been carried out according to plan; very likely this reference was to the 8 M/S and the T-boats and not, as the Naval Section suggested, 'to the activities of the 8 M/S Flotilla earlier'.[29]

Another reason why German confidence in the security of 'Schlüssel M' remained unimpaired is that just as much care was taken to protect the security of Ultra in documents liable to fall into their hands as in W/T transmissions. Among the British documents captured at St Nazaire were a minechart and instructions about German recognition signals for 26–31 March. In this case, shore-based radar could hardly have come into play as an alternative explanation for intelligence about minefields and swept channels; both items, as Hinsley pointed out, were derived from Enigma traffic. Nevertheless, despite several top-level investigations ordered by Hitler himself into the failure to detect the approach of the Chariot force, the Staff at Group West merely congratulated themselves on the accuracy of their own mine charts and continued to use their Enigmas with undiminished confidence. The Naval Staff (Operations) War Diary went to the length of adding an exclamation mark after the entry touching on recognition flares and morse signals![30]

It was not that the B-Dienst and Naval Group West were slow to fasten on to the possible implications of intercepted signals. On 31 July, for example, Devonport was heard to transmit a signal at 0244 to units at sea reporting that three E-boats out of Cherbourg would be passing Alderney at about 0315. What was remarkable about this signal was that the E-boats did not clear Cherbourg until 0250. Again, on 13 August, the B-Dienst pointed out to Group West that an MTB leader at sea had been advised by Dover at 2254 to wait for 'the main target'. That could only mean that he should refrain from attacking a decoy convoy then passing Gris Nez and wait instead for the valuable ore-carrier, *Grängesberg*, not scheduled to leave Boulogne for Dunkirk until 2330. In both cases, the local counter-intelligence branch of the Abwehr was called in at once to search for the agents assumed to be responsible.[31] Information obtained after the ambush and sinking of the *Turquoise* on 18/19 June – twenty-eight British prisoners were taken after the action – indicated that there had been foreknowledge of her sailing; but this was taken to mean espionage (*deutet auf Spionage*) or possibly aerial reconnaissance, since the convoy had left Le Havre in daylight. In December 1942, the Chief of German Naval Intelligence was commissioned to take sharper measures against the enemy's secret service which was reporting so promptly the departure of ships from Channel ports, measures that took the form of neutralizing local wireless transmitters and imposing intervals of interrupted telephone communications.[32]

Finally, the true context of such incidents involving intercepted British signals should be kept in mind, because what was involved should not be confused with some sort of German Ultra. Dover and other MTB bases transmitted their W/T signals in what almost amounted to self-evident groups which it was taken for granted the B-Dienst would be able to read. The Small Ships Code had earlier been used to communicate with MTBs but it was found that the time taken to decode signals at sea could mean the loss of tactical opportunities.[33] Supplying the enemy with a certain amount of gratuitous information about the range and accuracy of British radar was therefore the price paid to improve the chances of radar-controlled interception of E-boats off the south and east coasts or of convoys making their way through the Straits between Boulogne and Dunkirk. A 'B-Meldung' of 20/21 July, for example, came as a particularly unpleasant surprise by showing that radar on the Isle of Wight was tracking E-boats at a range of 37 nautical miles; until then, the best range reported had been 24 miles.[34] Such RN signals at least had an operational justification that the GAF broadcast of Freya plots lacked, since there was no alternative channel of communication. A milestone of sorts was passed early in September when four German minesweepers, with B-Dienst personnel embarked, navigated the Straits by means of the intercepted Dover plot of their progress every eight to ten minutes – the very role that Ruge had envisioned for *German* shore-based radar.[35] But the two incidents of 31 July and 13 August, which were by no means isolated cases, underscore the need for exceedingly careful handling of SI in Channel operations. There were various possible sources of intelligence other than SI for those signals to MTBs: Headache (intercepted E-boat VHF R/T transmissions), wireless traffic analysis, radar, aerial reconnaissance, even agents. Perhaps it is best simply to point out that on the night when SI provided advance warning (29/30 July), the B-Dienst intercepted no signals that might have been derived from it; whereas on the night when the suspicious signal was intercepted (30/31 July), the relevant decrypts did not reach the OIC until the forenoon of 1 August.[36]

In the confined limits of the Channel the air was constantly buzzing with W/T and R/T messages and all kinds of radar pulses. Some idea of the crescendo of lower grade wireless traffic during a major operation can be had from the report of an expert from GC&CS specially attached to 11 Group War Room on 19 August. Between them, the main RAF intercepting stations at Cheadle and West Kingsdown and their outstations collected eight separate categories of GAF R/T and W/T, ranging from shipping reconnaissance reports to Observer Corps transmissions.[37] The daily monitoring of GAF fighter 'point-to-point', providing data for periodic assessment of fighter movements and reinforcements, was done at West Kingsdown, Hawkinge and Beachy Head. In May, the RAF excluded wireless stations from the target list for fighter sweeps because the results of analysis of GAF traffic were more valuable than the destruction of the stations would have been. Not surprisingly, very little of its ephemeral raw material has survived, but RAF Y intelligence, on at least one occasion, played a decisive role in a combined operation. On 1/2 June 1942, six R-craft and four MGBs were on their way to St Valéry-sur-Somme (Operation Foxrock) when they were recalled on account of an intercepted GAF reconnaissance signal reporting their position 25 miles south of Dungeness.[38]

For GAF Y intelligence ('Funkhorchdienst') the situation is more problematical. Although the Y service provided almost three-quarters of its useful intelligence, according to one estimate, this achievement must be reconstructed where possible from oblique references and Allied sources, since the intercepts themselves have all been destroyed.[39]

The lack of records for the GAF Y service should not be allowed to convey an impression of a hopeless one-sidedness in favour of the RAF. In the spring of 1942, for example, GAF aircraft approaching the south coast of England were increasingly being warned by their controllers within a minute or two of RAF fighters being vectored to intercept them. One possible explanation was that Freya stations on the French coast were plotting the defensive patrols. In May, however, AI 4 was able to establish that the warnings were in fact the work of an exceptionally efficient Y station in the Cherbourg area which was monitoring RAF R/T. It and a similar station at Gris Nez were providing a 24-hour information service based on intercepted RAF R/T, and using daily changes in code and frequency. Not long afterwards, just before Yukon II, HQ Fighter Command warned the Groups about the danger of disclosing the scale, objective and date of forthcoming combined operations through poor R/T or W/T discipline. Special care was to be taken during the preliminary moves to advanced bases and during exercises and rehearsals. This order was apparently taken to heart, because the GAF reaction to Jubilee was noticeably sluggish; and its Y service later reported that the monitoring of RAF training and operational wireless traffic leading up to 19 August had disclosed nothing abnormal, which is to say that it failed to detect the movement of squadrons into the 11 Group area on 14 and 15 August.[40]

General Kessler's boast to David Irving about the breaking of British ciphers specifically mentioned naval ciphers, signifying the work of the German Navy's Y service or B-Dienst. The weekly B Berichte were not available for checking in 1963 but the bound originals were cleared for presentation at the National Archives in the late 1970s. B Bericht 34/42 for 17 to 23 August 1942 went out of its way to insist that there were no indications of a landing at Dieppe from wireless intercepts until the British broke wireless silence at 0528. The report blamed the GAF for failing to spot the assembly of small craft and shipping across the Channel before darkness fell on 18 August. In his published memoirs in 1981, Heinz Bonatz, who signed B Bericht 34/42 and who certainly would not have been slow in claiming any credit for his branch of naval intelligence had there been the slightest mention of Jubilee in signals intercepted before 19 August, lamented that aerial reconnaissance had probably not been flown. Three routine Dover signals were logged around midnight pointing to a possible target off South Foreland. Significantly enough, there was no mention of the two warning signals on Admiral's waveband from Portsmouth to *Calpe* and nothing at all about the encounter between Group 5 and the convoy from Boulogne. Hughes-Hallett, incidentally, had timed the breaking of wireless silence at 0540, when he ordered 2 Destroyer Division to 'Close inshore' and simultaneously received SGB 5's report of her disablement in the action with 'German trawlers'.[41]

One other item in B Bericht 34/42 covering 18/19 August is worth mentioning. The only incident detected by the B-Dienst during the night that was at all remarkable was the ramming and sinking of the 'MS *Golden Sunbeam*'. According to the intercepted signal, the *Golden Sunbeam* was in company with a damaged vessel and other boats ('weiteren Booten') and went down at 0400, or just after the action off Dieppe. 'Sunbeam', followed by consecutive numbers, happened to be the Return Passage Timetable code word for help messages for vessels withdrawing from Dieppe and 'Boote' could possibly have referred to landing craft. Indeed at 2 Defence Division, the report of this incident (B Meldung Nr 902) was entered in the war diary at 0810 on 19 August under the impression that it referred to the ramming of an enemy 'S boat' claimed earlier by UJ 1411. SO Group 5 was authorized to break wireless silence if he considered the landing of No 3 Commando to be in jeopardy, and he certainly tried to pass a signal to *Calpe* soon after the collision took place. His set, however, like his German counterpart's, was put out of action almost at once. Yet any speculation that the 'Sunbeam' signal might have come from Group 5 or had anything whatsoever to do with Jubilee is groundless. The *Golden Sunbeam* was a drifter, a veteran of the evacuation at Dunkirk, which sank after a collision off Dungeness.[42] The rest is simply, if rather weirdly, coincidental.

Naturally there could be no tactical Sigint until wireless silence was broken. Once that happened, the German Army Y service had a field-day, according to one operator whose letter describing his experiences on 19 August later fell into British hands.[43] But in the process of mounting combined operations, it was entirely possible that wireless traffic might have provided general intelligence about future British intentions. In his Battle Report, C-in-C West observed that there were three indications before Jubilee that an operation was imminent. First, there had been an increase in the number of agents parachuting into the Unoccupied Zone of France as well as in their wireless traffic. Secondly, Communists in Dieppe had begun to hoard food just before the raid. Obviously enough, neither of these manifestations had anything to do with a raid on Dieppe; but the third of von Rundstedt's alleged clues was ostensibly more telling: the British army abruptly introduced changes in wireless procedure (*Funkverfahren*) on 15 August, which greatly added to the difficulties of the German Y service, as reported in his weekly situation report on 17 August.[44] Changes of this description were generally taken to be a sign of an impending offensive. Could this possibly have been an ill-advised security measure that ran the risk of backfiring by drawing attention to the very operation it was intended to screen?

To start with, there are signs that Army wireless security on the Isle of Wight left something to be desired at the time of Rutter. On 9 July, Captain Kennard, the chief Security officer for the Isle of Wight, reported to GHQ Home Forces on his experiences since the Combined Operations operational and training base on the island had been set up. The Army's use of R/T before exercises and proposed operations had been quite appreciable, while at other times it had been negligible. According to Kennard, there had been a noticeable increase in GAF reconnaissance whenever the military use of R/T picked up. The Navy at Portsmouth carried on bogus R/T traffic on a daily basis, but the Army's failure to introduce a comparable programme was threatening to undermine the Navy's efforts. Kennard made no reference to the fighter-bomber attack on the LSIs on 7 July, though it is possible, perhaps, that a surge of wireless traffic in armoured units might have invited a pre-emptive strike. Jubilee, on the other hand, was mounted in a totally different way, there being no preliminary exercises like Yukon I and II, and the troops being assembled from their ordinary accommodation areas rather than concentrated for training on the island. And if there was a surge of military R/T sometime after 15 August, the Germans scarcely reacted as Kennard claimed they had done in June and July, because the GAF reduced rather than increased its reconnaissance effort between 15 and 18 August. In February 1944, Foreign Armies West admitted that a characteristic feature of Allied preparations to launch major amphibious operations had been their extraordinarily keen wireless discipline; in not a single case had Allied intentions been betrayed.[45]

Actually, there was no connection between the reasons for Kennard's misgivings and the changes abruptly introduced on 15 August, or indeed between the latter and Jubilee. The wireless changes were introduced by GHQ Home Forces without thought of impending operations. 'Experience has shown in various theatres of war', according to the order on 11 August, 'that the existing code name system and signal procedure are much too easily used by the enemy for identification purposes. A new system, known as the "Army Code Sign System", is accordingly being introduced to Home Forces, starting at 0001 hrs on 15 August.' The new system relied on 3-letter code signs that changed every 24 hours at midnight. 'Extracts', or lists giving the code signs for the week ahead, would thenceforth be prepared by higher formations and distributed down to battalion, squadron and battery levels. Code signs, never the names of units and formations, would invariably be used for R/T conversations and W/T messages in clear. A 'link-sign' system was to be introduced for use between HQs and subordinate HQs, and all regimental and staff officers would learn the Army Code Sign System and the link-sign procedure. The war diary of 2 Canadian division contains the Code Sign extract for the week beginning 15 August, but it was purely coincidental that this innovation came only three days before Jubilee.[46] All three of von Rundstedt's purported clues, in short, were not only *ex post facto* but bogus into the bargain.

The new British system was prompted by the capture of Rommel's field Sigint unit on 10 July 1942. It was discovered that Y intelligence had provided Rommel with a steady stream of information about 8 Army's order of battle and the tactical intentions of its commanders during operations.[47] In England, the German Army's Y service in the West enjoyed a reputation for efficiency at this stage of the war, before the proliferation of notional Army, Corps and divisional HQs in the UK for deception purposes. Unfortunately, that reputation is difficult to evaluate for the summer of 1942 for want of documentation. For instance, only seven Lageberichte West out of 35 issued by Foreign Armies West between 27 June and 22 August have been tracked down. Lagebericht West Nr 708 of 15 August failed to place 3 and 4 Commando correctly, although top priority was always given to tracing the movements of Commando and airborne units. The reorganization of the Canadian Army in England – 'kan. AOK' – carried out in April was mentioned only in Nr 675 of 27 June. While the Lageberichte contain intelligence from all sources, not just Sigint, presumably the Y service deserved credit for identifying various HQs involved in exercises in Northern Ireland, Norfolk and Scotland in July, but the available reports for June contain no mention of Yukon I and II or activities on the Isle of Wight. Twenty thousand Canadian troops in small parties appeared as part-time harvesters in Nr 696 of 18 July – source of information unspecified.[48]

Not that British intelligence about the German Army Order of Battle in Western Europe in 1942 was immeasurably superior to that provided by Foreign Armies West about formations in the UK. Ultra, for instance, may have been the source of misleading intelligence about the German division in Dieppe. The Hinsley Appendix was not very forthcoming about the mix-up between 302 ID and 110 ID, dismissing it as 'of less moment than those [other intelligence shortcomings] concerning the beaches and their defences'. But elsewhere in the Official History it was made clear that Sigint was one of the few reliable guides to operations on the Eastern Front, and that an important part of this intelligence concerned the German Army's order of battle in Russia. The OKW War Diary shows that 110 ID was the subject of top-level exchanges involving Hitler, the General Staff and Army Group Centre early in May. Instead of transferring the division to France there and then, as had been intended, Hitler decided on 3 May to leave it with 9 Army until the start of June. Although there was apparently not a great number of high-grade Army Enigma signals from the Eastern Front decrypted in 1942, there is at least a chance that news of this decision reached MI 14 from some Sigint source. That would seemingly explain why GHQ Home Forces Intelligence gave 11 June as the date for 110 ID's anticipated arrival in France. Hinsley also drew attention to the sporadic nature of Sigint concerning operations in Russia, which might explain why the subsequent decision not to transfer the division to France after all might have been missed. MI 14, for instance, reported to the CIGS on 13 July that the Grossdeutschland Infantry Regiment had not been identified since the previous October when it was operating in Russia, and complained on 24 August about the continued lack of reliable Order of Battle reports from the Eastern Front. At any rate, the continued presence of 110 ID in Russia was confirmed in the Martian Report of 26 August.[49] Here, for what it is worth, is another possible explanation of the 'mystery' of the Dieppe division's true identity to set beside that featuring the divisional emblem.

Such was the shortage of detailed intelligence for the planning of operations on the Continent in the spring of 1942 that the COS instructed the JIC to 'focus a searchlight' on Western Europe, one consequence of which was that the Combined Intelligence Section of GHQ Home Forces began to distribute the weekly Martian reports in June; these contained current intelligence from all sources, including carefully sanitized contributions from Ultra, that might be required for planning cross-Channel operations. The sharpest definition was achieved in covering naval and air matters where Sigint of all grades came into play in the course of operational activity. The German Army, by contrast, was able to rely on secure landlines, making basic Order of Battle intelligence difficult to collect. On 21 July, for instance, 712 ID was stated to have left France for Russia, but on 26 August it turned out that it had been occupying a stretch of the Belgian coast since mid-June; there was uncertainty all summer about 'the Cherbourg division' (320 ID). Starting in July, the Martian reports repeatedly forecast the departure of 10 Pz division for Russia, but it remained at Amiens until it departed for Tunisia in November. Report No 13 announced the arrival of 2 Pz in France on 26 August; No 14 corrected this on 9 September, 2 Pz never having left Russia. For months, 7 Fliegerdivision, one of the best known formations in the Wehrmacht, not to mention a potential offensive threat to the UK, was tagged as a 'parachute contingent' in 7 Army reserve in Normandy. Generally speaking, it was easier to establish which divisions had departed for Russia than to identify divisions involved in possible reciprocal movements to the West. Occasionally, the Germans themselves cleared up mysteries, as in the case of the 'Hermann Goering Brigade' known to be in Brittany. On 12 August, an appeal for volunteers in various specialties appeared in the German press. 'Like other crack Corps', the author of Martian Report No 17 drily commented on 23 September, 'the Goering brigade attaches no exaggerated importance to security.'[50]

Not surprisingly, then, the Detailed Military Plan made less of an impression on German Army staffs than did the navigational sketches on their naval counterparts. The staff at LXXXI Corps dismissed the intelligence about troops in the coastal zone as extraordinarily inaccurate, although they thought it noteworthy that the enemy was aware that 10 Pz was in 15 Army reserve, not, as its location might suggest, in Corps reserve, and that the Corps had been renumbered from XXXII.[51] Martian Report No 7 of 14 July had in fact gone into some detail about C-in-C West's reorganization carried out in June. The 'Rouen Corps' was now accepted to be at full establishment as part of extensive preparations to ensure that there would be enough Army and Corps staffs in the West to command the reinforcements that would pour in, in the event of an invasion – yet only on 30 September was the Corps commander named. In general, intelligence about the German Army in the West as presented in the Martian Reports, while more accurate and comprehensive than the data in the comparable Lageberichte West, was still largely a patchwork, subject to only incremental improvement, and often with areas of significant obscurity punctuated by arresting detail. General Kuntzen may not have been identified as LXXXI Corps commander for a number of weeks, but Field Hospital 1/613 at Amiens, one of the hospitals where wounded POWs from Dieppe would later be treated, was identified in July, when flat-footed members of 10 Pz were ordered to report there rather than to the overtaxed orthopaedic facility in Paris – source, once again, unidentified.[52]

There were occasions, finally, when Ultra failed to play the kind of role that had come to be expected of it. One of these was the air battle at Dieppe, which came as the culmination of a long strategic process. The policy of 'leaning forward into France' had been launched at the end of 1940 and was resumed in 1941, in the absence of a serious threat of daylight raids on England. The COS considered it important to maintain air superiority over the Channel in case the Germans once again threatened the south coast with invasion. Operations were intensified in mid-1941 when the urgency of meeting Russian demands for some show of offensive activity was agreed upon. The directive of 13 March 1942, under which the campaign was resumed, stressed the necessity of inflicting maximum wastage on the GAF. The Prime Minister assured the Chief of the Air Staff (CAS) that 'it pays us to lose plane for plane, and if you consider the "Circus" losses will come within this standard, it would be worthwhile'. The original objective of Sledgehammer, it should be remembered, was 'to force Germany to employ her air forces continuously in the West in conditions advantageous to ourselves'.[53] Unfortunately from the RAF's point of view, as the campaign developed, such optimistic expectations became increasingly difficult to realize, for technical and tactical reasons. By the end of 1941, the GAF had set up an efficient system of warning and control, which they continued to improve in the spring and summer of 1942, so that an error in gaining altitude at a rendezvous over southern England could lead to a surprise encounter over northern France at the moment of maximum disadvantage. The Fighter Group most involved, Leigh-Mallory's No 11 Group in the south-east, circulated a memorandum on 9 July admitting that operations were becoming increasingly costly. The GAF had pulled back from bases on the coast and given up attempts at interceptions over the sea; now they were content to wait until they could challenge penetrations beyond the coast from a tactically superior position. Worse still, the two Jagdgeschwader in France and Belgium (JG 2 and JG 26) had been steadily re-equipping with the FW 190, a superior fighter, with better rate of climb and heavier armament than either version of the Spitfire V with which 11 Group squadrons were currently equipped. The memorandum concluded that the Group faced two pressing requirements: to get the upper hand over the FW 190; and to neutralize by spoofing or jamming the GAF's early warning radar chain.[54]

Already, on 13 June, the Assistant Chief of the Air Staff (Ops) had instructed AOC-in-C Fighter Command that the policy governing the daylight offensive under the March directive must be adjusted with due regard to the technical superiority of the FW 190. There was a call for restraint in planning 'sweeps' or deeper penetrations into France, at least until the new Spitfire IX and Typhoon could be supplied in sufficient numbers to tilt the balance. In July, Fighter Command flew only half the offensive sorties it had flown in June. Whatever his outward enthusiasm for Rutter, the usually aggressive Leigh-Mallory was actually full of misgivings: quite apart from the superiority of the FW 190, his Group was constantly losing experienced pilots in overseas drafts; Dieppe was only just on the fringe of Fighter Command's 'advantageous zone'; his squadrons would be limited to 30-minute patrols over the town and committed to providing cover for fixed surface forces, so that they would be unable to disengage should the situation develop to their disadvantage. 'We can, I suggest, be well satisfied if our losses do not exceed, say, 60 to 70 pilots and 120 aircraft in the squadrons providing fighter cover', he wrote glumly to Sholto Douglas on 29 June. 'I do not quite know what you expect me to do about it', Douglas replied. 'I certainly do not propose to call the operation off. If I may say so, I think you are worrying too much about those possible casualties.' Perhaps this is the reason why, contrary to many published accounts, there is no mention in the Rutter or Jubilee Operation Orders or other planning documents of the inducement of an air battle as one of the raid's objectives. The RAF's objective was to protect naval and military forces against air attack and to delay the movement of enemy reinforcements.[55]

In the event, Leigh-Mallory was more than 'well pleased' with the outcome of the air battle on 19 August – 'the greatest air victory of the war', as he saw it. Following his cue, Churchill referred to Dieppe as a battle that Fighter Command wished they could repeat every week.[56] The GAF had indeed been forced to put up its greatest defensive effort over the Channel coast, including the unprecedented step of using night bombers in a day operation. AI 3b, responsible for evaluating all intelligence concerning the GAF's operations, order of battle, wastage and serviceability rates, reported that the 125 long-range bomber sorties on 19 August represented the commitment of all serviceable aircraft in France and the Low Countries except for the FW 200 Condors of KG 40 at Bordeaux. There was, however, always some question about German losses, which the *Daily Express* put as high as 273. According to a reliable source in Vichy, the Germans were openly admitting the loss of 170 aircraft, a figure Leigh-Mallory later laid claim to publicly. Leigh-Mallory was apparently convinced that his German opposite number would be replaced; when nothing of the kind happened, he decided that German losses on 19 August must have been concealed from higher authorities or somehow passed off as cumulative. On 23 August, Intelligence 3 at Fighter Command fixed German losses at 92 (including 43 bombers) and 39 probables, and calculated that the fighting strength of single-engined fighters in Holland, Belgium and northern France had been reduced from 312 to 150 as a result of aircraft destroyed or rendered unserviceable on 19 August. AI 3b at the Air Ministry later collated pilots' combat reports and camera gun films, intelligence from POWs from units engaged on 19 August who were subsequently shot down, and reports from the RAF Y service based on a study of W/T call signs for German bombers. Yet all this research, apparently, could not explain the fact that the German aircraft had been shot down over France or into the sea, often in circumstances where several pilots could plausibly claim the same kill. There was, however, a general assumption within the Air Ministry – expressed, for instance, by Air Commodore Whitworth Jones, Director of Fighter Operations – that the number of enemy aircraft destroyed was 'much greater' than the number claimed. It was also possible to put a favourable gloss on, for instance, the figures for fighters destroyed by taking into consideration the far greater number of RAF sorties (2,399 to c 600 for the GAF). In the end, there was no glossing over the fact that the GAF actually lost 48 aircraft, including 25 bombers. On the evening of 19 August, only 70 of the 230 single-engined fighters available in the morning were still serviceable, but by the following morning that number had risen to 194, thanks to overnight repairs and replacements (the last 18 FW 190s) from the aircraft-forwarding centre. The RAF lost 106 aircraft, the highest total for a single day so far in the war, and 81 pilots and aircrew killed or missing.[57]

In light of Peter Calvocoressi's observation in *Top Secret Ultra* that 'forty years later the mere thought of the range and extent of our knowledge of the Luftwaffe brings one up short', some explanation is called for as to why it was apparently necessary to wait until after the war to discover the true totals for enemy aircraft destroyed and damaged on 19 August 1942. Was the truth about German losses known but restricted to a small circle of the 'indoctrinated'? Actually, Calvocoressi was unintentionally misleading on this point, although it was necessary to wait to discover this until Hinsley pointed out in the Official History that Ultra intelligence about the GAF in the West was so rare in 1941 and 1942 as to be 'looked upon as an event of great note'. Ultra had already helped to sustain the costly and ultimately unproductive daylight fighter offensive before it reached a climax at Dieppe. When the GAF Enigma revealed on 6 July that the Germans had imposed flying restrictions in Russia and were encountering difficulties in supplying aircraft to North Africa, the Director of Intelligence at the Air Ministry accepted this as proof of the strain that Fighter Command was imposing on the enemy, whose losses would have an effect disproportionate to the number of aircraft involved. For the most part, Ultra's contribution seems to have been limited to providing intelligence about the movement of GAF units into and out of the western theatre of operations.[58]

A special section of the OIC handled air intelligence, working closely with AI in Hut 3 at Bletchley, because it was sometimes possible, according to Beesly, 'to gain insight into likely German naval intentions by deductions from GAF moves and plans'. The 'GAF Notes' included in OIC/SI summaries available at the PRO predicted and followed the movements of GAF formations whose role was a threat to shipping and combined operations – and, incidentally, helped to make up for the non-availability as yet of the GAF decrypts. For example, III/KG 26 first appeared as a Gruppe of torpedo-bombers after completion of the Aircraft Torpedo Conversion course at Grosseto 'about 15 July'. After its move to Rennes, it appeared in GAF Notes as having been 'misused' in a series of reprisal raids on Birmingham. Its first torpedo attack was launched against a convoy off the Scilly Isles on 3/4 August. The Notes for 23–29 August disclosed that some aircraft of III/KG 26 which had returned from northern Norway 'as recently as 21/8' moved back to Aalborg on 28 August: 'Comment: It is probable that all 20 a/c of this Gruppe are now under orders to return again to N Norway.' Also, although a few aircraft of III/KG 53 and II/KG 54 were identified during Jubilee, the Enigma confirmed that they were a handful left behind after the parent units left the West on 16 August. But neither Ultra nor lower grades of Sigint could provide reliable intelligence about GAF losses. AI 3b, after all, was among the Air Ministry's chief recipients of Ultra on the teleprinter from Bletchley Park.[59]

It seems reasonable to conclude that Jubilee would have gone ahead as planned even if Ultra had been disclosing how wasteful the daylight fighter offensive really was. According to the Air Historical Branch, the RAF claimed to have destroyed 197 enemy aircraft during the daylight fighter offensive in March, April, May and June 1942, at a cost of 259 to Fighter Command. The true German total was 58. 'Thus, far from the terms of the Prime Minister's "one-for-one" directive being complied with, we were, in fact, losing more than four times the number of aircraft the enemy were.' It should not be forgotten either that it was on the basis of those claims that ACAS(Ops) had called for an adjustment of the offensive on 13 June; or that the RAF was desperately short of Spitfires in the Western Desert, with only two flights in May 1942 as compared with Fighter Command's 59 squadrons at home.[60] A fully productive Ultra would no doubt have confirmed the RAF's role at Dieppe for what it was – the achievement of local air superiority against tactical odds and at considerable cost. For very different reasons, this was not a battle either the RAF or GAF would have wanted to repeat every week.

NOTES

1. Sir David Hunt, review of *The Ultra Secret*, *Times Literary Supplement* 13 Dec. 1974.
2. Ralph Bennett, *Ultra in the West* (London, 1979) and *Ultra and Mediterranean Strategy* (New York, 1989); Patrick Beesly, *Very Special Intelligence: The Story of the Admiralty's Operational Intelligence Centre 1939–1945* (London, 1977); Peter Calvocoressi, *Top Secret Ultra* (London, 1980). Bennett, 'Ultra and Some Command Decisions', *Journal of Contemporary History*, Vol. 16, No. 1, Jan. 1981, pp. 131, 132. See also Jürgen Rohwer and E. Jäckel (eds), *Die Funkaufklärung und ihre Rolle im Zweiten Weltkrieg* (Stuttgart, 1979).
3. Peter Calvocoressi, 'The Ultra Secrets of Station X', *Sunday Times*, 24 Nov. 1974.
4. James Rusbridger, 'The Sinking of the *Automedon* and the Capture of the *Nankin*', *Encounter*, Vol. LXIV, No. 5, May 1985, pp. 8–14; Phillip Knightley, *The Second Oldest Profession: The Spy as Bureaucrat, Patriot, Fantasist and Whore* (London, 1986), Ch. 5.
5. Hinsley, *British Intelligence*, Vol. 2, p. 192; Beesly, *Very Special Intelligence*, p. 227.
6. ADM 223/111, 1028 B/27 June 1942.
7. ADM 223/3, 'Increase of German Defence Measures, Western Area 1942', ZIP/ZG/173, 15 Aug. 1942.
8. DEFE 2/2, War Diary COHQ, 9 June 1942.
9. DEFE 2/650A, Operations Cancelled, 'C' to CCO, C/3177, 3 May 1943.
10. Lt. Col. Haines, *Ultra and the History of the United States Strategic Air Force in Europe vs the German Air Force* (Frederick, MD, 1980), p. 6. David Irving, letter to *Daily Telegraph*, 2 Sept. 1989. Ziegler, *Mountbatten*, pp. 190–91.
11. AIR 40/1781, AI 3(e) to ACAS(I), 2 June 1942; the decrypt in question was CX/MSS/1008/T25.
12. AIR 16/764; the note's distribution list was restricted to the three Jubilee Force Commanders and SOI, and it is to be found between two documents dated 11 and 13 Aug. 1942.
13. 'Führerbefehl vom 9 Juli 1942 betr Verlegung von Waffen-SS-Verbänden in den Bereich des OB West', OKW/WFSt 551 213/42 Gkdos Chefs, *KTB, OKW/WFSt*, II(2), pp. 1280–81; KTB, 302 ID, Ia Anlageheft C zum Kriegstagebuch Nr 3:

'Fernspruch von GenKdo LXXXI A K am 10.7.42, 2,35 Uhr'; 'Divisionsbefehl Nr 96 für den Küstenschutz', Abt Ia Tgb Nr 923/42 g, 7.7.42, T 315 2019.
14. DEFE 2/333, Fleming to CCO (attn Casa Maury), 21 Aug. 1942, with covering letter from DNI. Fleming wrote, 'All indications from Most Secret Sources point to surprise having been effected', and then changed it to read, 'No indications . . . surprise not having been effected'.
15. Hinsley *British Intelligence*, Vol. 2, Appendix 4: 'Enigma Keys Attacked by GC&CS up to mid-1943', pp. 658–68; Beesly, *Very Special Intelligence*, pp. 15, 23.
16. All decrypts in DEFE 2/187, which contains a thousand decrypts in the ZTPG series (68000–68999) received by the OIC, 16–19 Aug. 1942; the decrypts are bound according to their German Time of Origin (TOO), not according to their time of receipt by the Admiralty. They can of course be traced in German sources: for example, ZIP/ZTPG/68938 (TOO 0741) appears in KTB, 2 SichDiv, 19.8.42, T 1022 3552, as FT 0741/90/0820.
17. All decrypts in DEFE 3/187.
18. ADM 223/99, 'Admiralty Use of Special Intelligence in Naval Operations', by Cmdr G. Colpoys, pp. 16, 211. Beesly, *Very Special Intelligence*, p. 204.
19. Beesly, *Very Special Intelligence*, p. 227; Hinsley, *British Intelligence*, Vol. 2, pp. 190–91.
20. ADM 223/93, OIC/SI 135, 26 March 1942: 'The information that there were no permanent patrol positions in this area [Bruneval] is correct, but it is impossible to forecast when a patrol craft from Le Havre or ships on passage may be located in the vicinity.'
21. Hinsley, *British Intelligence*, Vol. 3 (1), pp. 286–7; DEFE 2/187, ZIP/ZTPG/69314, 29 July 1942, and ZIP/ZTPG/69418, 19 Aug. 1942; ADM 223/95, OIC/SI 319, 'Channel Activities – Noon 19 August to Noon 21 August', 23 Aug. 1942.
22. ADM 223/112, 1620A/18 Aug. 1942; ADM 223/95, OIC/SI 305A, 17 Aug. 1942, which updated OIC/SI 305, 17 Aug. 1942. For a comprehensive survey, see OIC/SI 312, 'Channel Activities – Noon 11 August to Noon 15 August', 19 Aug. 1942, and OIC/SI 313, 'Channel Activities Noon 15 August to Noon 17 August', 20 Aug. 1942. AIR 47/7, Photographic Reconnaissance, Vol. II (May 1941–Aug. 1945), pp. 99–100.
23. Beesly, *Very Special Intelligence*, p. 227.

24. ADM 223/94, OIC/SI 261, 'Possible Minelaying by M Class M/S', 10 July 1942; ADM 223/111, 1649B/11 July 1942.
25. ADM 223/111, 1217B/12 July 1942 was a message from C-in-C Portsmouth to the Admiralty: 'A movement of about 6 fairly large ships was reported by RD/F during the night 11/12 July leaving and returning to Cherbourg at 14 knots and passing the following positions . . .; 0059B/15 July 1942 and 1255B/15 July 1942 from the Admiralty to Cs-in-C Portsmouth and Plymouth and VA Dover gave the co-ordinates. ADM 223/3, ZIP/ZG/173, Appendix II, 'Minelaying 23/4–30/7/42', 15 Aug. 1942.
26. 'Ergänzender Bericht über Landungsunternehmen Dieppe', 1 Sk1 21 231/42 gKdos, n.d.: 'Der Gegner war ausserordentlich weitgehend bis in viele Einzelheiten über unser Küstenverteidigung unterrichtet.' T 1022 2363. KTB, MarGpW, 31.8.42, T 1022 3974.
27. 'Bericht des MarGpKdo West über Feindlandung bei Dieppe am 19.8.42 morgens'. Marinegruppenkommando West, B Nr Gkdos 3002/42 A1, 21.8.42, T 315 2019. ADM 223/4, 'German Naval "Y" Service', ZIP/ZG/223, 19 May 1943: the precise location of 'Naval D/F, Detachment Flanders' was not known but it was thought 'probable' that the intercepted Dover RDF traffic reports issued by 2 Defence Division were picked up by this organization: 'It is to be remembered that contrary to our procedure the duties of German D/F stations are not limited to the taking of D/F bearings but that they are provided with intelligence, and produce intelligence, as exploiting centres.'
28. The T-boats belonged to 3 T-boat Flotilla and were covered by a diversionary operation in the Straits of Dover (KTB, 2 SichDiv, 21.7.42, T 1022 3552). KTB, MarGpW: 'Kurzer Rückblick, 1 bis 31.7.42', and 15.8.42, T 1022 3974.
29. ADM 223/95, OIC/SI 312, 19 Aug. 1942. Whether the T-boats were plotted or not is impossible to say; four were known to have arrived at Le Havre on 16 July and to have left on 25 July for St Malo, two subsequently turning back to Le Havre (ADM 223/111, 1523B/20 July, 1633B/27 July, 1801B/27 July), but there is no mention of minelaying on their part. Reference to German Naval Summary (Channel Area) of 21 July from ZIP/ZG/173, Appendix II, 15 Aug. 1942.
30. KTB, MarGpW: 'Beurteilung Feindminentätigkeit in Zusammenhang mit Landungsoperationen', B Nr gKdos, 2053/42 A1, 9.5.42,

Anlage 6 zum KTB v 1.5–15.5.42, T 1022 3974; KTB, 1 Skl, Teil A, 28.3.42, T 1022 1672.
31. KTB, F d S, 31.7.42, T 1022 3253; KTB, MarGpW, 31.7.42, T 1022 3974; KTB, 2 SichDiv, 13.8.42, T 1022 3552.
32. KTB, BSW, 19.6.42, M 444. KTB, 1 Skl, Teil A, 26.12.42, T 1022 1678; KTB, 2 SichDiv, 20.5.42, T 1022 3552.
33. ADM 219/13, 'Interception of E Boats by Coastal Forces', Directorate of Naval Operational Studies, Report 14/42, 29 Dec. 1942.
34. KTB, 1 Skl, Teil A, 21.7.42, T 1022 1675; KTB, 2 SichDiv: 'Lagebeurteilung am 15.7.1942', T 1022 3552.
35. KTB, BSW, 12/13.9.42, M 444: The commander of 2 Minesweeping Flotilla reported, 'Bei Kanalmarsch laufende gutliegende engl Ortungen, die zur Navigation besser ausnutzbar als deutsche.' See also, Hugo Heydel, 'Sicherungskräfte im Westraum 1939–1943', *Marine Rundschau*, 51 Jahrgang, Heft 2/195, 55–65.
36. DEFE 3/184, ZIP/ZTPG/65180, ZIP/ZTPG/65219, which reached the OIC at 0927 and 1050 respectively on 1 Aug. 1942.
37. AIR 40/2239, "Y" Intelligence for Combined Operations', including the report of 30 Aug. 1942 by A.W. Bonsall (GC&CS), who was present in the War Room at No. 11 Group during Jubilee.
38. AIR 40/1781, DCAS Bottomley to Leigh-Mallory, 6 May 1942. DEFE 2/215, Operation Foxrock, Commodore Commanding, Dover, to Admiralty, 9 June 1942.
39. Wilhelm Haenske, 'Die Luftnachrichtentruppe 1944 im Westen', *Wehrkunde*, IV, Nr 3, 1955, 91–8.
40. AIR 40/318, 'Enemy Reporting of RAF Fighter Patrols', Group Captain Cadell, AI 4(b), 4 May 1942. AIR 16/760, Operation Rutter: Minutes of Meeting, 15 June 1942.
41. Roskill, 'The Dieppe Raid and the Question of German Foreknowledge: A Study in Historical Responsibility', *Royal United Services Institution Journal*, CIX, No. 633, 1964, pp. 27–36. RG 457, German Navy Reports of Intercepted Radio Messages, B Berichte 1942, Band II, Nr 26–52: Oberkommando der Kriegsmarine, Skl/Chef MND III, 3345/42, Nr 34/42, gKdos, 27.8.42. Bonatz, *Seekrieg im Äther*, p. 163. WO 106/4197, Jubilee Naval Force Commander's Covering Letter, No. NFJ/0221/92, 30 Aug. 1942.
42. Peter C. Smith, *Hold the Narrow Sea: Naval Warfare in the English Channel 1939–1945* (Ashbourne and Annapolis, 1984), p. 199.

43. DEFE 2/324, War Office MI 8/B 2340 to COHQ, 21 March 1943: 'Among various papers found near Maruth, apparently, were some personal letters from a Lt Klatcis in Amsterdam to a Lt Wesel in Libya: "The Dieppe business bucked up our chaps (Y service) enormously . . ."'
44. Oberbefehlshaber West (Oberkommando Heeresgruppe D), 'Gefechtsbericht über Feindlandung bei und beiderseits Dieppe am 19.8.1942', Ia 2550/42 gKdos, 3.9.42, T 78 311. 'Lagebeurteilung durch Ob West', Ia Nr 2436/42 gKdos, 17.8.42, T 78 311.
45. DEFE 2/550, Captain L. Kennard to Major W. Gibson, GHQ Home Forces, 9 July 1942. The OKW War Diary lists a photographic reconnaissance of Southampton at 0800 on 17 August, which detected five 'passenger ships' (LSIs?), but this was not part of an intensified search and does not seem to have been followed up (*KTB, OKW(WFSt)*, II (1), p. 603); British sources claim that no overland reconnaissance was flown on that day. Generalstabes des Heeres, Abt Fremde Heere West III, 'Einzelnachrichten des Ic-Dienstes West Nr 28', Nr 1763/44 geh, 18.2.44, T 78 450.
46. RG 24, C 3, 13,746, War Diary, G Branch, HQ 2 Cdn Inf Div, Oct. 1941–Aug. 1942: 2DS(G)3–4–12, 'Notes on the use of the Army Code Sign System', 11 Aug. 1942.
47. Hans-Otto Behrendt, *Rommel's Intelligence in the Desert Campaign 1941–1943* (London, 1985), pp. 168–86.
48. Handakten Etzdorf betreffend Lageberichte West Nr 511–712, vom 10 September 1941 bis 19 August 1942. Etzdorf's collection is by no means complete, there being a gap, for example, between No. 598 of 19 Jan. 1942 and No. 671 of 21 June 1942.
49. Hinsley, *British Intelligence*, Vol. 2, Appendix 13: 'Intelligence Before and During the Dieppe Raid', p. 700, pp. 69-70; *KTB, OKW(WFSt)*, II (1), p. 334; WO 208/3573, MI 14 Weekly Summary for CIGS, 18 March 1941–25 June 1944, 13 July 1942, 24 Aug. 1942; *KTB, OKW(WFSt)*, II (1), 3.5.42. WO 219/1934, Martian Report No. 13, 26 Aug. 1942.
50. Hinsley, *British Intelligence*, Vol. 3 (II), Appendix 2, 'The Martian Reports', pp. 753–55. WO 219/1933, Martian Reports Nos 1–12 (12 June–19 Aug. 1942).
51. KTB, 302 ID, Ia Anlageheft 'Dieppe' zum Kriegstagebuch Nr 3: 'Gefechts- und Erfahrungsbericht über den englischen Angriff auf D am 19.8.42', Ia 640/42 geh Kdos, 25.8.42, T 315 2019.
52. WO 219/1934, Martian Report No. 18, 30 Sept. 1942; RG 331,

SHAEF, G-2, Operational Intelligence Sub-Division, Martian Reports, Box 34 (Nos 79–98), Report No. 86, 4 March 1944.
53. AIR 41/49, 'Air Defence of Great Britain, Vol. V, The Struggle for Air Supremacy (January 1942–May 1945)', p. 103.
54. AIR 16/638, 'Reports of Air Fighting in No. 11 Group', 11G/500/13/DSASO, 9 July 1942.
55. AIR 16/760, Leigh-Mallory to Sholto Douglas, 11G/S 500/86, 29 June 1942; Sholto Douglas to Leigh-Mallory, FC/S 28669, 30 June 1942. AIR 8/895, Leigh-Mallory to CAS, Notes for Rutter, 2 July 1942.
56. AIR 6/748, 'The RAF at Dieppe', by Air Marshal T.L. Leigh-Mallory CB DSO. Churchill quoted in *The Times*, 9 Sept. 1942.
57. *Daily Express*, 21 Aug. 1942; Leigh-Mallory quoted in *The Times*, 26 Feb. 1943; a report – FC/S.30073/Ops – claiming that the Germans had admitted the loss of 170 aircraft was forwarded by HQ Fighter Command to Groups on 7 Sept. 1942 (AIR 20/5186). AIR 24/562, Notes and Appreciations No. 41, Intelligence 3, HQ Fighter Command, 23 Aug. 1942; AIR 40/1784, Dieppe Operations 19.8.42, AI 3(b), 17 Sept. 1942. AIR 20/5186, Air Commodore Whitworth Jones to AOC-in-C, Fighter Command, 4 Sept. 1942. AL 2646, Luftflottenkommando 3, 'Erfahrungsbericht Nr 1', Führ Abt/Ia op Nr 8655/42 gKdos, 28.8.42.
58. Calvocoressi, *Top Secret Ultra*, p. 133; Hinsley, *British Intelligence*, Vol. 2, p. 236.
59. ADM 223/94, OIC/SI 268, 'GAF Notes 12–18 July', n.d.; ADM 223/95, OIC/SI 329, 'GAF Notes 23–29 August', 29 Aug. 1942, and OIC/SI 316, 'GAF Notes 16–22 August', 23 Aug. 1942. For the 'misuse' of experienced aircrew on the Baedeker raids, see AIR 41/49, 'Air Defence of Great Britain', Vol. V, p. 83. AIR 40/1781, Sqn Ldr Lee, AI 3(b) to DDI 3, 4 April 1942: 'Estimates made of enemy losses and effort are both subject to a very wide margin of error on the present way of computation.'
60. AIR 41/49, 'Air Defence of Great Britain', Vol. V, pp. 114, 132–4.

6

Dieppe and D-Day

While he was still GOC-in-C South Eastern Command in November 1941, the future Field Marshal Montgomery attempted to summarize the 'Lessons Learnt' so far in the war. The first (of 14) dealt with what he termed the stage management of battle. 'Against an enemy as good as the German', he wrote, 'it is difficult, almost impossible, for a formation or unit to recover if it is put into battle badly in the first place.'[1] Montgomery was thinking in orthodox terms and was to give his words expression at El Alamein a year later. Yet the supreme example of the set-piece operation during the Second World War was the combined operation. Opposed amphibious landings lent themselves to stage management because they were subject to a process of advanced planning in detail and doctrinal refinement from one operation to the next. Jubilee and Overlord – the latter the greatest set-piece in history, let alone military history – have in addition always been assumed to share a special relationship purely as cross-Channel operations. In keeping with the theatrical imagery, Jubilee has enjoyed a reputation as the 'Rehearsal for Invasion,' as the trial run whose lessons gave new meaning to what Montgomery characteristically called 'a good kick-off'.

Naturally, those, like Mountbatten, Crerar and Hughes-Hallett, whose attitude towards Jubilee was defensive and justificatory, readily adopted the notion of an operational rehearsal and the direct relationship it implied between the two operations. Crerar put it forcefully in an address to 3 Canadian Infantry division on the eve of their assault on Juno beach:

The plan, the preparations, the method and technique which will be employed are based on knowledge and experience bought and paid for by 2 Canadian division at Dieppe. The contribution of that hazardous operation cannot be overestimated. It will prove to have been the essential prelude to our forthcoming and final success.

Mountbatten apparently found any failure to appreciate this point of view exasperating. 'I do hope', he wrote to the retired Canadian general, C. Churchill Mann, in 1973, 'that the "Dieppe boys" will have at last understood that without their valiant efforts we could never have had Overlord.' Ironically enough, Wallace Reyburn, war correspondent, survivor of the raid, and author of the wartime best-seller *Rehearsal for Invasion* (1943), later repudiated the concept altogether and became outspokenly sceptical about the value of the 'Lessons Learnt' at Dieppe and critical of their cost in lives. He and Hughes-Hallett – acting as a proxy for Mountbatten – celebrated the 25th anniversary by becoming involved in a rancorous controversy in the *Sunday Telegraph*, which was reproduced in over 70 Canadian newspapers and came to an end with Hughes-Hallett threatening legal action unless given the last word.[2] Even outside this inflamed context, the question of Jubilee's relationship to Overlord is difficult enough to bring into focus, but at least there are now enough documents available to put some of the more ambitious claims to the test. Take those very disparate aspects of the raid's operational context already touched upon in some detail: deception and radar.

If Jubilee made a major contribution to the success of the cover and deception plans for Overlord, then a reassessment of the sacrifice made on the beaches of Dieppe would indeed be in order. Overlord was the decisive operation of the war against Germany for the Western Allies, whose two greatest advantages in 1944 were air power and the power to deceive. Cover is mentioned only in passing in the Combined Report on Jubilee under 'Security' in 'The Lessons in Detail'. Section C (ix) reads:

> The early production of a 'cover' plan for the forces engaged, including not only their training but their moves, is an urgent necessity.

This is a reference to defensive cover; strategic deception was too sensitive a subject to be included in a printed document with a fairly wide distribution, although all forms of tactical feints and diversion were endorsed to confuse the enemy during the actual landing. Defensive cover for an operation like Jubilee would have meant the appointment of an I(b) staff officer to the planning staff to detect any inconsistencies between the cover and the operational plans and to keep the staff informed about the general state of security and leakages.[3] That the mild recommendation above can be adduced as a 'Lesson' of any particular significance for Overlord may be questioned, since it was inconceivable even in 1942 that the cross-Channel invasion would be risked without the most carefully planned and elaborate cover and deception. But there is always the chance that unacknowledged lessons for the practice of deception were drawn at the time; or that Jubilee might have produced some kind of delayed effect that turned out to be useful in planning the invasion or in pushing the Germans along the road towards self-deception. Since the success of deception in 1944 was not widely known until after Masterman's book in 1972, claims staked on behalf of Jubilee would have been correspondingly delayed.

Actually, there is nothing to suggest that Rutter and Jubilee were planned with cover and deception for future operations in mind. A German analysis, published in *Wehrwissenschaftliche Rundschau* in 1978, of Allied deception covering the choice of assault area for the invasion failed even to mention Dieppe. Apparently stung by this omission, General C. Churchill Mann made a feeble attempt to inject an element of contrivance into the operation by making out that the Operation Orders captured by the Germans were a deliberate plant. The Germans were supposedly misled into interpreting the raid as some sort of preliminary to an invasion of the Pas de Calais. Fortitude South, the most important of Overlord's cover plans, did indeed pose the threat of an invasion of that stretch of the French coast, but its origins cannot be traced to Jubilee, despite Churchill Mann's best efforts. No-one knew, in the summer of 1942, where the invasion would come ashore and therefore where the cover area would be; apart from that, Dieppe is in Normandy, not the Pas de Calais, which for Allied purposes stretched from the Somme to Gravelines. And the ill-advisability of taking complete copies of Operation Orders ashore was one of the more specific of the security lessons drawn from Jubilee. In fact this far-fetched notion would not rate a mention but for the facts that Lieutenant-Colonel Churchill Mann was GSO 1, 2 Canadian division during the planning of Rutter, was involved with the planning of Jubilee after promotion to Brigadier General Staff 1 Canadian Corps on 13 July, and indeed served as deputy Military Commander on board *Fernie* during the operation.[4] Jubilee certainly did encourage the Germans to look for further raids on the French coast later on in 1942, thereby increasing their nervousness in the West. But even here it is going too far to invest the operation with a preconceived role in strategic deception for Torch, as Cave Brown did. The raid could not have been intended 'to give teeth to Overthrow', since the decision to mount Jubilee preceded the decision to invade French North Africa by about ten days.

Mountbatten stated in 1973 that 'Dieppe unexpectedly, but most fortunately, became one of the great deception operations of the war'.[5] The Germans supposedly misread its lessons and went on strengthening their port defences and garrisons in the belief that a full-scale invasion could not be sustained for long across open beaches. There is a good deal of truth to this, inasmuch as the German defensive system was organized around the principle of denying the Allies the use of a deep-water port in working order for as long as possible. The Naval Operations Staff pointed out on 5 September 1942 that the first requirement for a successful invasion was a port capable of providing logistical support for a force of some 20 divisions; only Cherbourg, Le Havre and the Hook of Holland had the unloading capacity to meet this requirement, the Channel ports lacking size and being easy to block or sabotage. Apparent confirmation came later from large-scale landings in the Mediterranean, leading Foreign Armies West to the conclusion in February 1944 that Allied operations conformed to a pattern, the most important feature of which was proximity to a 'Grosshafen' like Naples. But the importance of defending major ports did not originate with the Dieppe raid: it was a matter of strategic first principles. Apart from the question of supply, for instance, landing too far from a major port would force the invaders to sacrifice penetration inland to the need to advance laterally along the coast, thereby exposing a shallow beachhead to counter-attacks aimed at segmenting it. The principle had been unerringly fixed from the very first naval and military appreciations of the invasion threat; it underlay von Rundstedt's reorganization of defences begun in May; and it had been one of the themes of Hitler's discourse on the Atlantic Wall on 13 August. The naval document of 5 September never mentioned the quite recent raid, either as the source of any new lessons or the confirmation of well-established ones.[6]

What about the possibility that Jubilee served a useful purpose by enabling the Allies to push the 'story' for the sake of deception that because of their experience at Dieppe they could not risk landing very far from a major port? After all, Mountbatten did use the term 'deception operation'. This question is difficult to answer definitively for lack of access to the full range of sources, but enough can be deduced from the Official History and available documents to suggest that such a 'story' could well have been counter-productive. There was only one possible choice of cover area for an invasion in Normandy, and the beauty of deciding to land west of the Seine was the opportunity that presented itself of developing the known attractions of the Pas de Calais as viewed by the Germans. This appeal, resting on such considerations as the shortest Channel crossing, maximum fighter cover, promising lines of strategic exploitation, and so forth, had been embedded in the German military mind from the beginning. It was part and parcel of the comprehensive assessment drawn up for Zeitzler's benefit by his predecessor as Chief of Staff to C-in-C West in March 1942; von Rundstedt himself never deviated much from the view that the invasion would take place somewhere near the Somme estuary; and Foreign Armies West were of the opinion in April 1944 'nach wie vor' that the main attack would come across the Straits of Dover.[7] Thus the two predominant themes in German anti-invasion planning were the obvious advantages for the Allies of the Pas de Calais and the necessity of their capturing a major port as quickly as possible.

Not the least of the success of Fortitude South lay in glossing over the incompatibility of these two themes, for the most serious drawback to the Pas de Calais as cover area for Overlord was its distance from adequate port facilities. To have stressed the idea that the Allies could not land too far from a major port would have been to risk diverting German attention towards the Bay of the Seine. Hitler showed signs in April and May 1944 of expecting an invasion there, precisely because the bay was flanked by two first-class ports. In brief, Jubilee did not present the planners of Fortitude South with the invaluable asset implied by Mountbatten; and the success of the deception did not lie in directing German attention to beaches closer to Cherbourg or Le Havre but instead further afield to the coast well to the east of the Seine. As it happened, the German Navy considered the invasion beaches unsuitable for a landing for various reasons, including the lack of a decent port and the presence of underwater reefs, strong currents and, at some points, cliffs.[8] Surprise is not always the result of deception.

For Jubilee to have played a decisive role in deceiving or misleading the enemy, an altogether different set-up is required, starting with a successful raid and the capture of Dieppe. Deception would have come into play had the Allies then continued to give every indication through special means and other channels that they were counting on repeating this success on a larger scale by storming Cherbourg or Le Havre, while all the time preparing to land in the Bay of the Seine. The truth of the matter is more prosaic: the raid was not a success and the Allies landed where they did out of necessity rather than choice.

Secondly, radar. In the Appreciation and Outline Plan for Overlord (COS(43)416(O)) of 30 July 1943, the seriousness of the threat to tactical surprise posed by the number and variety of German radar installations was spelled out at some length. Coverage was complete between Haugesund in Norway and the Spanish frontier; between Dieppe and the Channel Islands alone there were by then some 25 to 30 long-range installations with considerable overlap in coverage. The destruction by air attack of the majority of them would be no guarantee of neutralizing the early warning system in the invasion area because of the mobility of some types of equipment and the ease with which others could be replaced. Fortunately, there was some possibility of technical interference or confusing the Germans by feints and diversionary attacks.[9] At first glance, it seems far-fetched that Jack Nissen's exploits at Dieppe or any of the other radar-related events of that day could possibly have contributed to overcoming this challenge. F 28 was not captured, which ruled out any answers to queries about IFF or anti-jamming devices, and neither could anything have been learned about the effectiveness of various methods of spoofing and jamming because none was tried. Besides, other types of equipment came into service soon after August 1942, such as the Mammut (or Hoarding); on the eve of D-Day, for example, there were two Freyas, two Giant Würzburgs, one standard Würzburg and a Hoarding arrayed at Dieppe. Nissen also made an easily disproved claim that GEE was used for surface navigation 'for the first time' on D-Day as a result of recommendations he made after Dieppe. In fact there were experiments with GEE even before Rutter, and already a year before the invasion it was being referred to as 'a well-tried method'. How much better does his claim that Jubilee provided 'the impetus' for developing devices that were the key to the successful electronic counter-measures used on D-Day stand up to a documentary test?[10]

There is more to the Nissen brief than mindless affirmation that lack of jamming for Jubilee demonstrated the urgency of providing it for Overlord. His great achievement was to confirm that the Freya had become a 'precision' radar. The frequent swivelling and stopping of the aerial array during the air battle raging overhead made it clear that the operator was plotting individual targets. 'Beyond doubt, Tait [Air Commodore and Director-General of Signals and Radar at the Air Ministry] was right. Freya was a precision radar. That was what I had come to find out.' Tait, according to Nissen, had been involved in a controversy with Lindemann over the correct target for counter-measures: Freya or Seetakt? Once Nissen's report settled the issue in favour of the Freya, Tait ordered the mass production of the Mandrel 120 MHz jamming equipment, which could provide enough 'white noise' on the Freya's frequency to lay down a curtain that severely curtailed accuracy and range. This highly secret device was given 'some flying tests' but saw only limited service so as to preserve technological surprise for D-Day.[11]

Mandrel was actually one of two radar counter-measures under development by TRE in 1942, the other being codenamed Moonshine. At first both were thought to have an application to daylight fighter operations, since Mandrel could screen the approach of a sweep to within 30 miles of the coast while Moonshine could simulate a large formation elsewhere, using a mere handful of aircraft; they could also prove to be useful for combined operations in the Channel. Moonshine was tested on 9 August, when nine Defiant aircraft flying in a pattern in mid-Channel provoked a strong GAF fighter reaction in the Cherbourg area. Oculist made it possible to verify the results by intercepting and reading the W/T plots transmitted by the Auderville and three other Freyas. Moonshine's lifespan was more or less as Nissen described Mandrel's – limited operational employment before being withdrawn in October and then used on D-Day.[12]

In Mandrel's case, ground sites were chosen at Dover and Hastings in May and performance trials were held on 8 July; on 28 August, the Air Ministry advised the manufacturer at Wembley that delivery of a supply of transmitters was 'a matter of very great urgency'. This urgency, however, had nothing to do with combined operations or the Freya's precision rating but rather with the bomber offensive. Mounting casualties in Bomber Command, it was becoming clear, were the consequence of improvements in 'the enemy's radio aids to night defences'. The introduction of night fighter zones and the removal of searchlights from those zones suggested that the Germans had developed a technique analogous to the British GCI/AI system using Freyas and Giant Würzburgs. Until then, Bomber Command had been opposed to the use of counter-measures, but this policy was now to be reversed. Ground Mandrel stations on the English coast would be directed at Freyas on the French and Belgian coasts; airborne Mandrel carried by aircraft in the bomber stream would interfere with inland GCI Freyas. Airborne Mandrel, Bomber Command's first offensive jammer, was introduced on 5/6 December 1942.[13] Moreover, from the start Mandrel was given far more than limited use, largely to exploit to the utmost its operational life expectancy of about six months, by when it was assumed the enemy would counter with frequency changes or more radical measures to make the GCI system less dependent on the Freya. In March 1943, R.V. Jones compiled a report based on Ultra of the effects achieved by ground and airborne Mandrel against the German '120 Mc/s RDF'. In May, the Germans were reported to be designing a Mandrel homing device, using sets retrieved from downed bombers. On 8 August, HQ Bomber Command drew attention to the necessity ('a most urgent matter') of providing more airborne Mandrel to cover the extended frequencies (120–150 MHz) on which the Freyas were then operating; eventually, the range covered from 75 to 200 MHz.[14]

Mandrel survived to play its part in the electronic battle on D-Day, alongside Window, Moonshine, Grocer, Cigar, Jostle, Carpet, Boozer etc. Airborne Mandrel, for instance, was used as part of the counter-measures in support of the airborne landings. There was also, it should never be forgotten, a costly campaign of direct air action against 'all installations which could not be jammed, or which could report on shipping or be used to control batteries, or which were located in areas where their presence might endanger our airborne forces . . .' Would the Mandrel story have been very different had Jubilee never taken place? Hardly. Jubilee lacked the decisive impact it would have had by demonstrating the need for such a device in the first place; or if it had been the occasion of Mandrel's first operational use. The Germans had indeed modified their Freyas to improve D/F bearing in the spring of 1942, but it apparently made no technical difference to Mandrel whether the change to 'precision' radar had been made or not. It is simply impossible to accept Nissen's claim that Mandrel would 'never have been designed, developed and manufactured without the traumatic blow we received at Dieppe'. Or, indeed, that attention would have been switched to the Seetakt (which was already being jammed) if he had never got ashore. In fact it is legitimate to ask if Nissen provided any information not available, then or not long afterwards, from other sources. His statement that it was no longer possible in 1942 to eavesdrop on German W/T plotting, as it had been in 1941 'before the restoration of the German telephone system', implies that only by cutting the telephone cables leading from F 28 was he able to force the station to resort to W/T. There is some evidence to suggest that reliance on W/T had declined somewhat since Jones's report in January – which commented on this inclination during periods of low activity. According to the GC&CS report on Jubilee of 30 August, 'usually RDF decodes are only available from the French coastal area west of Fécamp'. On the other hand, J.R. Robinson (1991) rather convincingly challenges the claim that Nissen had cut the vital cables at all; they had been cut earlier by bombing, although in his opinion this made little difference since passing plots by W/T was still the normal procedure at the time of the raid. Whatever the truth of the matter, it seems possible that F 28 had not recently been heard from in as great a volume as in 1941. On the other hand, Oculist was still in business for the Freyas west of Fécamp (presumably already modified), as the results of the Moonshine test on 9 August prove; and R.V. Jones apparently saw no reason at the time to write a supplement to his January report on the Freya and Würzburg or to make even a passing reference in his memoirs (1978) to the value of any special radar intelligence gained as a result of Jubilee. Tait was quoted in 1967: 'He [Nissen] didn't get very much. But he did bring back some information that was helpful in developing counter-measures'. This measured assessment has a more convincing ring than the hyperbolical claims later made by Nissen, and apparently Tait as well, to the effect that D-Day would have been impossible but for the radar information brought back from Dieppe. Quite apart from anything else, of course, a large volume of technical information about German radar was collected from POWs and documents in the Middle East in 1943.[15] In short, if Jubilee provided keys to unlock the secrets of success on D-Day, it is going too far to say that any of them turned either the radar or the deception lock.

One of the greatest of the many ironies surrounding the Dieppe raid is that it was widely believed at the time that it had provided the key to the most important lock of all. The first requirement for the success of Overlord, according to the Appreciation and Outline Plan, was the reduction of the GAF fighter force in the West, 'which, unless checked and reduced, may reach such formidable proportions as to render an amphibious assault out of the question'. Leigh-Mallory had written to Mountbatten on 22 August 1942 that the most important consequence of the raid had been to force the enemy to fight in the air to the limit of his resources. Every effort should therefore be made to launch another raid or series of raids to induce further air fighting and so progressively wear the GAF down. A landing on a two- or three-mile stretch of coast should provoke the enemy into another violent reaction in the air, especially if an armoured force were landed to wreak havoc inland. It was only a matter of putting the troops ashore in the correct way at a weak point rather than at a heavily defended port. A series of such raids could 'lead up to the destruction of the Luftwaffe'; and the Germans would be forced to commit their mobile reserves and so present the ground attack squadrons with the kind of targets not seen at Dieppe. This proposal was well received by CCO and endorsed by the COS, but the acute shortage of landing craft in the UK after meeting the requirements for Torch quickly reduced what sounded like a watered-down version of Imperator to a token landing; and it was just as quickly realized that a landing by a small party of troops would present the enemy with the opportunity to score another propaganda victory by representing it as a failed attempt at invasion. There remained an outright feint by a naval force, in which case Leigh-Mallory was concerned to get the GAF off the ground as fast as possible. 'If the Germans take as long to get going as they did in the Dieppe show', he wrote to Sholto Douglas on 28 September, 'it would be apparent to them that the landing was a hoax and they would undoubtedly take very little interest in the operation.'[16]

The assumption that further raiding could bring on another 'advantageous' air battle was widely shared at COHQ, although operations like Jubilee were not considered in the general interest at the War Office and GHQ Home Forces. 'There remains . . . a school of thought', one Brigadier wrote to another in March 1943, 'which considers that Dieppe was a howling success because we were able to bring the GAF into the air and shoot down 200 German aircraft at the expense of 100 of our own (plus, of course, 5,000-odd good soldiers).' Francis Williams, Controller of Press and Censorship at the Ministry of Information, was sensitive to the danger of giving the impression, especially in Canada, 'that we cheerfully accept all those losses purely for the sake of an air battle'. Planning at COHQ went ahead anyway for feints or raids as soon as the landing craft situation improved. Leigh-Mallory's proposal evolved into Operation Aflame, a feint off the Somme estuary by an impressive naval force whose surprise appearance at dawn, calculated to set off an air battle, was originally to have been launched between 4 and 16 October. In other words, this was the operation that would have given 'teeth to Overthrow', if, that is, Torch had not been postponed until November and if Aflame itself had ever got off the ground instead of being victimized by the weather. Aflame left a lasting impression, however, as witness references in the documents to operations 'of the Aflame type'. Among the chief agreements at the Casablanca conference was one to launch raids in 1943 with the object of inflicting attritional losses on the GAF; and, as already mentioned, COSSAC's directive in April included the preparation of an elaborate camouflage and deception scheme over the summer, featuring at least one feint by an amphibious force to try to replicate the air victory at Dieppe. The culmination of much detailed planning by COSSAC's staff was Starkey in August and September 1943 – a large-scale deception that came to a head with the sailing of an 'invasion force' on 9 September, and in which the Germans did indeed 'take very little interest'. Starkey had no chance of impressing the enemy without the display of a substantially greater number of landing craft on the south coast than had been visible at the time of Jubilee, but this was impossible because of the demands made by another Mediterranean operation – this time Husky on 10 July.[17]

It was excusable that the air battle at Dieppe was mistaken for an RAF victory; less excusable was the failure to appreciate before September 1943 that attempts to repeat it by using raiding forces as bait or by pure deception led to a strategic dead-end. It should have been clear that the GAF had over-reacted at Dieppe and would not do so again unless convinced that the Allies were attempting to secure a permanent foothold on the Continent. As the daylight offensive demonstrated, the Germans were in the position of being able to refuse battle until they had sized up the situation. Yet they were expected not to be deterred by what the RAF believed to be very considerable losses from putting up another maximum effort at the drop of a hat. Secondly, Leigh-Mallory, and to a lesser extent Mountbatten, showed a remarkable indifference to political and strategic realities after Jubilee. These realities were summed up in December by Lieutenant-Commander A.N.P. de Costobadie of COHQ but should have been apparent for the previous three months. De Costobadie agreed that a raid could produce another air battle but doubted that even the most powerful of raids would achieve any success as such, now that German defences were so highly developed within Fighter Command's advantageous zone. 'In this case', he wrote, 'it is most doubtful if we are justified in suggesting a raid which if it failed would produce a public outcry.'[18]

Little acknowledgement has been made of the lessons learned from Starkey for the invasion, yet they were considerable. Starkey, indeed, provided the missing spark of inspiration for what was shaping up as a fairly predictable cover plan for Overlord. The (pre-Starkey) Appreciation and Outline Plan called for a diversion against the Pas de Calais in divisional strength on about D-14, possibly as the final stage of the general air plan to reduce the GAF fighter force. Starkey showed that the GAF was unlikely to be brought to battle without actually carrying out the landing. But putting a division ashore somewhere near Calais or Dunkirk without a follow-up would soon have been unmasked as an attempt to tie down German reserves, and so would not maintain the continuity of threat to the Pas de Calais necessary until the main invasion force was firmly established in Normandy. Besides, there would not be enough landing craft to mount such an operation except at the expense of the main assault two weeks later. The effort put into diversions should always more than repay the loss of effort thereby incurred in the main attack. The Germans expected an invasion on a broad front, with plenty of diversionary operations to conceal the 'Schwerpunkt'; Foreign Armies West, for example, cited a raid north of the Garigliano on 17/18 January 1944 in their analysis of the Anzio landings on 21/22 January as confirmation that the Allies would carry out this sort of diversion before or during a *Grosslandung*. Therefore, if this very early version of the cover plan had stood, von Rundstedt would have felt free to reinforce his ground forces south of the Pas de Calais soon after D-14. The 'elegance' of Fortitude South, according to Masterman, was that it prolonged the threat of a major assault and follow-up by the notional 1 US Army Group across the narrowest part of the Channel for weeks after the supposedly diversionary landings in Normandy had taken place. This stratagem was based on the enemy's inflated estimate of the Allied Order of Battle in the UK and implemented in various ways, including special means and the shifting of the centre of gravity of visible invasion preparations in the south of England to the east.[19] Thus it was Starkey, not Jubilee, that 'unexpectedly, but most fortunately', yielded the key lesson for cross-Channel strategic deception in 1944.

To bring the relationship between Jubilee and Overlord into sharper focus, one or two undeniably obvious points should be made. First of all, no mere raid could in any literal sense have been a rehearsal for an operation of the sheer magnitude and complexity of the cross-Channel invasion in 1944. Overlord's basic prerequisite was nothing less than the reorganization of the Army and Metropolitan RAF for offensive operations on the Continent and the assembly of massive US air and ground forces in the UK. Mountbatten observed that a raid on Dieppe could teach nothing about the problem of maintenance over open beaches: that, to put it mildly, was the least of it. He might have continued *ad infinitum* about all the other aspects of the invasion, from the massive preliminary air campaign to decontamination arrangements in case the Germans used poison gas, which had no counterpart in raids, even in miniature. Raids required none of the elaborate administrative planning that was the main preoccupation of the Combined Commanders in 1942, much of which was later found to be useful by COSSAC's staff; and the trickiest stage of a raid was the withdrawal and re-embarkation – the 'worst possible case' contingency for invasion planners. It would have taken a raid on a far greater scale than any attempted to yield lessons about the marshalling, embarkation and sailing of an invasion force. Starkey, the simulated invasion, included a large-scale movement exercise taking troops up to the point of embarkation, plus the imposition of Restricted Areas on the south coast, minesweeping in the Channel in daylight in the hope of testing the effect of bombing on the accuracy of German coastal batteries, and other features that made it far more of a dry run than Jubilee could possibly have been.[20] Strictly speaking, the lessons learned from a raid like Jubilee that might be expected to have applied directly to an invasion were those dealing with its sharp end – the assault.

Secondly, there is the complication that rather a long hiatus separated Jubilee from Overlord; if the former is to be loosely considered a 'rehearsal' for the D-Day assault (Operation Neptune), it was certainly not a dress rehearsal. Moreover, during the hiatus several large-scale amphibious operations took place in the Mediterranean, including Husky which involved two armies and used almost as many landing craft as Neptune. Yet there was a school of thought led by Crerar which held that none of the 'varied information' acquired in the Mediterranean conflicted with the lessons learned at Dieppe or materially added to them. Hughes-Hallett was always of the opinion that operations in the Mediterranean had little relevance to those in the Channel because of the totally different weather and navigational conditions. COSSAC himself pointed out to the COS in July 1943 that in many ways the recent invasion of Sicily and the future cross-Channel invasion could not be more dissimilar. Husky involved the use of bases along an extended coastline for a converging attack on an island; for Overlord, it would be necessary to launch an assault from an island against an extended coastline.[21]

Sweeping generalizations like these always invite fine-tuning. There were of course innumerable lessons from operations in the Mediterranean, many of them analysed and circulated in COHQ Bulletins. COSSAC himself, Lieutenant-General Morgan, did go on to point out in his statement to the COS that Husky had provided 'invaluable experience' for the detailed planning of Overlord. Clearly the Combined Chiefs of Staff did not agree with Hughes-Hallett about the primacy of familiarity with Channel conditions over operational experience of large-scale operations in the Mediterranean: otherwise Admiral Sir Charles Little, C-in-C Portsmouth, would have remained C-in-C Expeditionary Force instead of being replaced by Admiral Sir Bertram Ramsay (Eastern Naval Task Force commander for Husky) at the end of October 1943. And among Crerar's 'varied information' compiled in the Mediterranean were data for the large-scale use of airborne forces. COSSAC included parachute and glider operations in his invasion plans from their inception – airborne forces were earmarked for the capture of Caen in one of the earliest versions – but Jubilee had nothing to offer beyond a general endorsement of their use, so that only after evaluation of their performance in Sicily could serious planning of their role in Overlord begin. Then again, the introduction of new equipment such as the LST and DUKW led to revised estimates of the efficiency of Beach Groups in maintaining forces for prolonged periods under favourable weather conditions. Another welcome surprise with Husky and Avalanche was the unexpected effectiveness of direct and indirect naval fire support; at Salerno (Avalanche) there was no preliminary bombardment to preserve surprise, which was lost anyway, and the operation was largely saved by the guns of the Allied navies.[22]

It nevertheless remains true that the cross-Channel invasion was quite rightly viewed from the earliest days as a specialized operation which would require careful planning and preparation over a prolonged period. Given coastal defences far exceeding in strength any encountered in the Mediterranean, it was accepted that success in the all-important assault would depend on speed and special technique and equipment; moreover, there was the constant threat that, if the correct basic technique was not evolved, those defences might be strengthened to such a degree that methods which promised success in 1943 would no longer be adequate when the time came. If that happened, the assault would have to be carried out largely by airborne forces or some way found to extend air cover beyond its limited range; or, in the worst possible case, the intention to invade might have to be abandoned altogether. Churchill was quoted in *The Times* on 4 August 1944 to the effect that Dieppe had laid down 'definite though broad' conclusions affecting the invasion. If this is so, the best period to search for guidelines would appear to be during the formative stages of invasion planning in the late spring and summer of 1943, when enough time had passed for Jubilee's lessons to sink in and before they were overlaid by lessons learned elsewhere.

What effect did Jubilee have on the choice of location and tactical method for the D-Day assault? The invasion beaches were chosen first, before it had been decided whether to land under cover of darkness or in daylight. There had been long discussions in 1942 about the selection of the assault area for Roundup between those like CCO who favoured Normandy and the advocates of the Pas de Calais – usually Army and RAF spokesmen impressed by the advantage of the shortest Channel crossing under maximum fighter cover. This question so bedevilled the Combined Commanders' staff, according to Morgan, that they produced no operational plan 'worthy of the name'. But after the original Roundup plan was found not to be feasible, they produced a paper 'Selection of Assault Areas for a Major Operation in North-West Europe,' CC(42)108 – which concluded that the *only possible* choice was the Caen beaches together with beaches on the eastern side of the Cotentin peninsula, provided they were assaulted simultaneously.[23] COSSAC's staff made the same selection towards the end of June 1943 – 'to assault the Norman beaches about Bayeux' – with a similar sense of inevitability, giving the planning process an impetus and a realism that had been missing until then.

Jubilee's contribution to this decision, according to Hughes-Hallett, was to prove the impracticability of capturing a port with unloading facilities intact. A more accurate conclusion would surely have been that it proved the impracticability of capturing a port at all, since the Germans sank no blockships and blew up none of the harbour installations at Dieppe. The unpublished history of Combined Operations (1956) in the PRO agrees: 'The main lesson learned was that a frontal assault on a defended port was not practicable and none was carried out subsequently to Dieppe.' This was not, incidentally, listed with the other Lessons Learnt in the Combined Report; and the origins of the Mulberry harbours, a concept with a number of historical roots, can be fixed, realistically speaking, in June 1943 – too late in the day to be directly ascribed to Dieppe.[24] Two questions arise. Was there ever a serious intention before Jubilee to plan an invasion to be dependent on the prompt capture of a port or ports in working order? If so, how reasonable is it to suppose that planning would have gone forward based on this premise but for Jubilee?

The answer to the first question is affirmative. One feature of invasion planning in its very early stages was an assault on a broad front. In May 1942, the Combined Commanders were thinking along the lines of landings at points between Calais and the Cherbourg peninsula, from which beachheads would be expanded to capture aerodromes in the Pas de Calais and the ports of Cherbourg and Le Havre; once linked up, the beachheads would form a zone extending from Calais to St Quentin, along the Oise to its junction with the Seine, and from there to the south-west corner of the Cherbourg peninsula. In June, the Prime Minister only half-jokingly called for Roundup to be based on 'magnitude, simultaneity and violence', with far-flung diversions in Denmark, Holland or Belgium, plus the early capture of aerodromes and no fewer than four ports. The Outline Plan for Roundup approved by the COS on 30 June featured landings in the Pas de Calais, Seine North and Seine South sectors, and at Cherbourg.[25] Earlier in the year, according to his memoirs, Hughes-Hallett was approached by a Brigadier from GHQ Home Forces who was working on what must have been one of the very first invasion plans. This operation was to take the form of simultaneous attacks in divisional strength on four or five ports between Boulogne and St Brieux; in each case, flanking attacks were to be launched in brigade strength while the third brigade was held back in floating reserve. One point still to be decided was whether the reserve brigade should be used to reinforce the flank attacks or launched in a frontal attack timed to coincide with the arrival of one or both of the pincers. The Brigadier had got wind of Rutter and asked Hughes-Hallett if it would be possible to plan the forthcoming raid so as to test those tactical concepts. After approval by VCCO, a COHQ plan for flanking attacks on Dieppe was drawn up, but at a meeting on 18 April a rival GHQ Home Forces plan incorporating a frontal attack was adopted instead. So, at least, Hughes-Hallett would have it. Stacey described the idea of an original COHQ plan for flanking attacks on Dieppe as a 'myth', largely out of concern to refute the rumour that it had been the Canadians who had insisted on the frontal attack. In point of fact, Dieppe was initially chosen as a target by a COHQ committee in conjunction with BGS (Plans), GHQ Home Forces, and the same BGS (Plans) and the Deputy Chief of Staff, Home Forces, came out in favour of a frontal attack rather than COHQ's Scheme B for flanking assaults at the meeting on 18 April, further explaining their reasons on 25 April. As a result, the concept of a frontal assault preceded by heavy bombing was built into the Rutter Outline Plan approved by the COS on 13 May. It remains a minor mystery why the murderous raid on St Nazaire did not leave more of an impression than it apparently did. In any case, Sledgehammer East, in the second half of July, was amended to incorporate a frontal attack on Boulogne, and the COS had been particularly interested in May in German preparations to use blockships and other methods to obstruct ports and basins in the event of a landing. 'You will remember', a former senior member of COHQ wrote to Mountbatten in 1973, 'that the Prime Minister, as well as the COS, was insistent that there should be a reconnaissance in force across the Channel that autumn to test out the feasibility of capturing a Channel port for Operation Overlord. As there was no time to plan a new operation you told them that the only way of achieving this was for the Dieppe operation to be mounted again.'[26]

An invasion on a wide front based on the capture of four or five ports of modest capacity had little prospect of success, even as an operation of opportunity, in the spring of 1942; it would have been open to defeat in detail and impossible to cover from the air. The only possible justification for taking such a proposal at all seriously was the still primitive state of German coastal defences. The Abercrombie and Bristle raiders, for example, had encountered no serious obstacles on shore, apart from some easily cut wire, and next to no resistance. Only one casualty was sustained (wounded by friendly fire) during Abercrombie; out of a Bristle force of some 250, the toll was one killed, two missing and none wounded. The earliest Martian reports in June 1942 went to some length to draw attention to the fact that the enemy had yet to take some of the most basic anti-invasion precautions. Such defences as there were had been thrown together to guard against local raids: wire was fairly common at beach exits but nowhere very formidable; anti-tank obstacles were rare away from the relatively well-defended ports; the mining of beaches had only just been reported west of Le Havre; and there were no underwater obstacles or beach scaffolding, no blockships in readiness at any of the Channel ports. The most noticeable defensive activity disclosed by aerial reconnaissance was the obstruction of aerodromes close to the coast and the laying of underground telephone cable from town to town along the coast to fill in gaps in the prewar system, which radiated outward from Paris. Whatever its shortcomings revealed during Jubilee, RAF PR coverage was far more comprehensive and systematic than anything the GAF was able to achieve over England. It picked out, for example, thick belts of wire being laid in the rear and on the flanks of coastal towns like Dieppe, Le Tréport and Fécamp; in some cases, this wire might have been in place unnoticed for some time but grass growing in the spring and left untouched by grazing animals had betrayed its location.[27]

Actually, this general state of unpreparedness had already begun to change after May 1942 – too recently to register in the earliest Martian Reports? – following von Rundstedt's tour of inspection and radical reorganization of his defences around a limited number of strongpoints, complete with concrete shelters for personnel, weapons and ammunition. The Dieppe defences, like those of every other coastal town, were significantly strengthened between the end of May and the raid. In June and July 1942, the first serious anti-tank obstacles began to appear, a feature of German coastal defences that was pushed ahead with great urgency after Dieppe, together with the upgrading of anti-tank guns from 37 mm to 50 mm and 75 mm.[28] As a result of the strenuous work of fortification noted in von Rundstedt's weekly reports for the autumn and winter of 1942–43, by the following spring the case against launching either a frontal or a pinching-out attack against Cherbourg, Le Havre or Antwerp, the only three ports with the capacity to support an invasion, was conclusive.

The Overlord Appreciation and Outline Plan spelled this out in some detail. Le Havre, for example, was well prepared for demolition and protected by interlocking arcs of fire from coastal batteries that made a direct approach to the mouth of the Seine 'a matter of extreme hazard'; as for flanking attacks, there were no suitable beaches immediately to the north or east of the port from which to launch sufficiently powerful forces to overcome its landward defences. Even if taken, Le Havre could not function until the batteries on the opposite bank of the Seine had been captured or neutralized; similarly, the good landing beaches west of Trouville were within range of artillery on the Havre peninsula. Quite apart from that, simultaneous landings in the 'North and South Seine sectors' would be difficult to consolidate because the first bridge was well upstream at Rouen and the river below there was unsuitable for bridging by pontoon. There were equally compelling strategic and geographical reasons why Cherbourg and Antwerp could not be captured quickly; and the Pas de Calais beaches and harbours had long since been written off as lacking the capacity to sustain a force large enough to expand a beachhead to include either Le Havre or Antwerp.[29] Thus, not only would the Germans have turned all ports into fortresses even if the raid had never taken place, but no invasion plan based on the quick capture of one of the major ports by frontal attack or *coup de main* could have survived scrutiny much beyond the autumn of 1942 – even if the raid had been a success. An outdated or poorly conceived experiment often yields irrelevant data.

Tactical method, or doctrine, was a lot more complex than the choice of the assault area. It was one thing, clearly, for the Combined Commanders to deduce where the invasion would have to be launched: quite another to anticipate how the assault would be carried out. For the D-Day assault, tactical doctrine was the responsibility of COSSAC, who received his directive from the Combined Chiefs in April but did not produce a doctrine for the assault until weeks after the completion of the Overlord Appreciation and Outline Plan at the end of July. Any evaluation of Jubilee's contribution to this doctrine must begin with the 'Lessons' spelled out in the COHQ Combined Report in October 1942, which were published separately and given the widest possible circulation by Mountbatten – at the Casablanca conference, for example. The Report was delayed by a printing bottleneck and the preoccupation of the staff at Richmond Terrace with the planning of Torch, not, as Mountbatten was careful to point out, by attempts to doctor it; it contained 'all our latest available knowledge' and was 'devastatingly frank'.[30] Nine months later, in June 1943, Mountbatten convened the Rattle conference at Largs, Ayrshire (HMS *Warren*), 'to study the Combined Operations problems in Overlord', partly out of disappointment at the relatively minor and advisory contribution his staff had so far been asked to make to the invasion planning. Enough time had presumably passed by then for the Lessons to have established their validity or not. Rattle was attended by COSSAC and his Principal Staff Officers, Cs-in-C Portsmouth, Home Forces and Fighter Command, with a large number of their staffs, and by Lieutenant-General J. Devers, representing HQ ETOUSA. Rattle took place before Husky and before there was an operational plan for the invasion, giving its papers and deliberations an abstract quality at times, but in the opinion of many of its participants the invasion could never have taken place without it. After Rattle and a number of exercises in August and September, COSSAC's staff at Norfolk House put the final pieces together, and the tactical doctrine for Neptune was given its first demonstration by 3 Canadian division in Exercise Pirate in mid-October.[31]

The Jubilee Lessons were a mixed bag, even as a set of general guidelines. It is impossible to read them without being reminded of how much basic thinking had been done at the tactical level *before* Jubilee. They covered every aspect of a combined operation, from planning and training, to security and the briefing of the troops, to the need for a higher standard of aircraft recognition, to control and communications. They were listed in 'Summarised Form' and then enlarged upon in 'The Lessons in Detail'. An emphatic statement in summary – 'The importance and necessity of using smoke cannot be over-emphasized . . .' – could end up subject to elaborate qualifications in detail. Other Lessons dealing with critical aspects of an assault, such as its timing, were surprisingly open-ended. Assaults would have to be carefully timed, but there were 'certain conditions' varying from one operation to the next that would decide whether to land at dawn or dusk, in darkness or in daylight. One reason for this equivocalness, remarkably enough, was that the Lessons were drawn up with future raids in mind, as well as the cross-Channel invasion, and the requirements for the former were much more variable than those for the latter. Some useful lessons known to have been learned at Dieppe were unaccountably omitted, such as the value of having a forward fighter controller on one of the HQ ships. Another was that a low-category formation such as 302 ID was liable to be just as effective as a high-category one when used in a static role in well-constructed defences. One of the chief Lessons, according to Hinsley, concerned the interpretation of low-grade Sigint before forwarding the results to Commands; since this process required a knowledge of high-grade Sigint, it could best be done by the Air Section at GC&CS. For obvious reasons, this too found no place in the Report. Nothing was said either about the wisdom, or lack of it, of remounting an operation after it had been cancelled.[32]

Pride of place went to a Lesson whose obviousness enraged Reyburn and whose validity Hughes-Hallett and Mountbatten claimed to have been fully aware of well before Jubilee – the need for 'overwhelming' fire support in a frontal assault on a well defended position, including close support in the final stage of a landing. Fire support would be delivered by heavy and medium naval bombardment, strafing by cannon-fighters, and high-level bombing; special shallow draught vessels or even some sort of mobile forts would cover the assault formations during the critical period after the lifting of the preliminary barrage and before the formations were able to deploy their own artillery on land. Given enough fire support, so it seemed, virtually no frontal assault was impossible, including those launched without surprise against a heavily defended target. This confident ring was only slightly softened by the subsequent admission that reliance could still be put on surprise in the absence of overwhelming fire support, in which case the assault should be directed at the flanks of a fortress rather than frontally against it; if this Lesson was meant to apply only to raids, it did not say so.

Secondly, Jubilee had been planned to test the role of armour in an amphibious assault, the first waterproofed Churchill tanks landing in daylight on the heels of the infantry on Red and White beach. But the tanks fell victim to anti-tank guns or failed to overcome the various obstacles in their path from the water's edge to the streets leading into town. Roberts sensibly decided that the employment of tanks in a raid of short duration was not to be recommended. Hughes-Hallett went further and decreed that Jubilee had shown that the landing of tanks in the first wave of an assault was not a practicable proposition. One of the most categorical of the Lessons decreed that henceforth tanks should not be landed until the anti-tank defences had been cleared or destroyed; LCTs, too, which had drawn intense fire at Dieppe, should not be delayed on the beach after unloading their tanks by the need to disembark sappers with their demolition charges. The Germans, incidentally, drew exactly the same conclusion: the Allies were nothing if not cautious and methodical and they would not land tanks until the infantry had won them room for manoeuvre to exploit the beachhead. This Lesson was really a reversion to conventional doctrine calling for the infantry to be landed from small, fast LCAs in the dark and on a rising tide, followed by tanks and other heavy equipment from the slower LCTs at first light; by then, the beach should have been cleared of obstacles and protected from anti-tank and direct artillery fire. 'It is assumed', ran the Outline Plan for Hadrian early in 1943, 'the infantry will be landed before dawn and the tanks and SP guns as soon as there is light enough for them to manoeuvre.' All the large-scale landings in the Mediterranean touched down well before dawn. It is perhaps also worth pointing out that the Army had lost interest in developing a heavy 'assault' tank; according to a General Staff policy recommendation on 26 February 1943, the tactics favoured for an attack on a position strongly defended by minefields and wire, as at El Alamein, called for the infantry and sappers first to clear gaps under cover of a heavy bombardment for the new, all-purpose medium tanks.[33]

Several of the Lessons advocated flexibility in the planning and carrying out of operations, another piece of conventional military wisdom but considered doubly applicable to combined operations. Dieppe had illustrated the unprofitability of using reserves to reinforce hold-ups rather than to exploit what success there was. A distinction was made between committing maximum forces all at once in a raid and switching reserves from one point to another in a large-scale operation. An invasion assault should be spread over several beaches, depending on how wide a frontage could be controlled and the strength of support available; also, the minimum force necessary should be committed to the assault and as many troops as possible held behind to back them up. Even if a flexible plan had been drawn up for the Dieppe raid, it was added, it could not have been used because of the low level of training among landing craft crews: a state of affairs that could only be remedied by the creation of permanent naval assault forces to train with the Army formations they would actually land.

The Lessons were broad enough, in short, to encompass several assault models. If 3 Canadian division had been about to land on Juno beach at, say, 0200 to seize a beachhead for the landing of a follow-up force of tanks at first light, Crerar's remarks to his troops on the eve of D-Day would have been just as appropriate. At the Rattle conference, there were surprisingly few overt references to Dieppe, despite the presence of Leigh-Mallory, Hughes-Hallett, McNaughton and, of course, Mountbatten. Occasionally, the impact of events the previous August can be discerned between the lines. Hughes-Hallett, for example, stressed the 'serious menace' posed by the enemy's normal coastal convoy traffic; even if only a few ships were encountered on passage, the whole invasion expedition could be thrown into confusion. There was a reminder that every effort should be made to supplement intelligence from PR by tapping all possible sources such as agents. A COHQ paper – 'Pros and Cons of an Assault in Daylight or Darkness' – expressed a slight preference for landing in the dark so as to avoid observed small arms fire, and a certain reluctance to land at first light. Dawn was the traditional time for defences to stand-to and any delay 'might bring it [the assault] into the category of a day operation, without the measures peculiar to a day operation being taken'. That of course was exactly what had happened at Puits. One thing, however, that the conference papers did reflect quite unambiguously was an intense concentration on what had come to be called 'the problem of the assault'.[34]

This emphasis was so pronounced as to suggest a fixation with the first few hours of the invasion, even allowing for Morgan's acknowledgment in his Appreciation that 'the crux of the operation' would be resisting a counter-attack by German reserves rather than breaking through the coastal crust. Although Morgan scarcely mentioned Dieppe in *Overturn to Overlord* (1950), and explicit references to it in the documents became increasingly infrequent as the range and complexity of planning increased, it is hard to imagine where else such a fixation might have come from. Significantly enough, it was Morgan's American deputy, General R. Barker, who pointed out during Rattle that too much stress was being put on the strength of the coastal crust of defences; the Germans were holding 'a very wide frontage' and there was every chance of getting ashore, particularly if 'the first landings were carried out at first light'. This remark made no impression at all on his British audience and the conference quickly moved on to the next paper. But the theme of excessive preoccupation with the landing itself, according to Max Hastings, was to become 'a commonplace' of historical studies of the campaign in Normandy. It was picked up, for example, by the distinguished American historian Russell Weigley, who described the Allied planners as so totally immersed in solving the problems of the assault that they neglected 'the maze of troubles awaiting behind the French shore'. The British in particular became entangled with 'military gadgetry of varying utility', because of their caution and inability to sustain heavy casualties. Nor was Ramsay left in any doubt that the extreme detail of the naval orders for Neptune 'was foreign to the practice of the US Navy'.[35] Thus an extreme case could be put together whereby Jubilee contributed to the comparative neglect of planning for the follow-up and build-up stages of Overlord, and yet by some cruel twist also failed to provide novel lessons of decisive value for the assault.

Neither point carries conviction. First of all, the assault was the one part of the invasion that lent itself to advance planning in detail and which the Allies could not afford to have go wrong. The strength of German coastal defences had increased several-fold between June 1942 and June 1943: there was no telling at that juncture how strong they might be by June 1944. Whatever General Barker might say, the invasion still looked decidedly risky as reflected in the Rattle documents and even the Appreciation and Outline Plan. There would be no chance of strategic surprise and little enough of tactical surprise. There was every reason to expect the assembly areas on the south coast to be heavily attacked by the GAF. The seaborne assault was to be on a three-divisional front, without the landing on the eastern Cotentin beaches that the Combined Commanders considered essential. There were serious reservations about the power of the preliminary naval and air bombardment to neutralize, let alone destroy, coastal batteries except by chance. Such a bombardment had never been tried before 1943 and next to nothing was known about the effect of near misses on the occupants of pillboxes, casemated batteries, and other positions protected by 6 feet or more of reinforced concrete. Indeed a proposal had been made in March 1943 to subject Alderney to an experimental bombardment and then land a Commando to monitor the results.[36] In February 1944, moreover, Ultra confirmed that the Germans had decided to defeat the invasion on the beaches, rather than as heretofore by armoured counter-attack. The coast, according to Rommel, was to be held 'absolutely'. Large numbers of underwater obstacles appeared for the first time that month, taking Allied Intelligence by surprise. To suppose, as Hughes-Hallett and Crerar evidently did, that this was a tactical mistake inspired by German confidence as a result of Dieppe that an invasion could be held on the beaches, is quite mistaken, because this reversal of strategy was based instead on experience of the landings in the Mediterranean. Max Hastings suggested that the comparative neglect by the invasion planners of the post-assault phases counted only for the afternoon of D-Day and 7 and 8 June; after that most of the Allies' problems were the result not of shortcomings in planning but of the superior fighting quality of the German Army – in itself, it might perhaps be added, reason enough to be relieved at the attention lavished on the assault.[37] Thus it is perhaps legitimate to claim for Jubilee a pervasive but largely unspoken effect on the awareness of the planners in the form of a salutary reminder of the irremediable consequences of a badly planned assault.

Secondly, COHQ was not the only authority to go to school

on Jubilee. Not all analysts came to the conclusion that tanks had no place in the initial assault. Lieutenant Colonel G.C. Reeves of the Department of Tank Design considered that the landing of tanks on a heavily defended coastline was still an operational possibility, if likely to be a costly one. Major B. Sucharov, Royal Canadian Engineers, pointed out in his report on Jubilee that if tanks were to be landed under heavy fire they should be equipped with mechanical aids for placing charges, laying ramps and so forth, and that the sappers with whom they would have to work would themselves require armoured protection. Within eight days of the raid, Lieutenant J.J. Denovan had already produced a drawing of an engineer tank, which eventually evolved into the various versions of the Armoured Vehicle Royal Engineers (AVRE). A War Office request for the development of such a tank was sent to the Ministry of Supply on 2 February 1943. There already was a General Staff requirement for a Duplex Drive (DD) or swimming tank before Jubilee; on 3 August, for example, the Director of Fighting Vehicles at the War Office assigned a high priority to the production of the Straussler flotation gear for the new Sherman medium tank.[38]

This is the most significant prompt leading from Dieppe to D-Day. But it is difficult to document and easy to oversimplify. Liddell Hart, for example, stated that Dieppe demonstrated the need for specially trained assault troops and armoured equipment. 'After this lesson had come to be more fully realized . . .' Major-General P.C.S. Hobart was appointed to command 79 Armoured division in March 1943. Hobart undertook the job of developing DD, AVRE and Canal Defence Light (CDL) tanks and training their crews. But who 'realized' the lesson? There was evidently no clear-cut decision by a technical sub-committee or War Office directorate to ignore the COHQ verdict about the place of tanks in the assault. Indeed, at a meeting on 30 June 1943, the Director of the Royal Armoured Corps referred to it as though it were still valid, and a COHQ representative reported that there was no requirement for DD tanks in combined operations, though some interest had been shown in submersible ones.[39]

The testing of tactical doctrine on schemes and exercises was not the responsibility of 79 Armoured division but of 1 Corps. 1 Corps had been Morgan's command before his appointment as

COSSAC and it moved to Scotland in the spring of 1943 to begin work as 'the assault Corps'. The Corps War Diary indicates only limited progress before Rattle; at a Study Period in mid-June, the best time for H-Hour was thought to be at first light after a night with a three-quarter moon; the assault would use AVREs and be carried out under cover of an area smoke plan, in full acceptance of the limiting effect of smoke on the efficiency of various forms of fire support. At Largs, the Corps commander, Lieutenant-General G.C. Bucknall, spoke up in favour of as much fire support as possible but went on to argue that the secret of success would be the landing of an integrated assault group including AVREs and DDs, which implied an assault in daylight. Approaching the coast at night and landing infantry just before first light to secure the beach for the tanks and SP artillery had almost always failed; if the beach was 'resolutely held', he added, the infantry would be pinned down and the landing of heavy equipment could turn into a very risky undertaking, as at Dieppe.

Bucknall's presentation took Hughes-Hallett by surprise – 'a dramatic intervention'. If this proposal was accepted by COSSAC it would mean 'revolutionary changes' in the training of the naval assault forces. Landing craft could cross the Channel only on LSIs or in company with other craft of the same type and speed, and assembling a composite group made up of various types would be impossible in the dark. This was the clinching argument for operating in daylight from the lowering point in, for it was too late now to design and produce an all-purpose landing craft capable of carrying tanks, SP artillery and a platoon of infantry. It was also evident of course that the AVREs needed daylight to do their work efficiently. In August, accordingly, Hughes-Hallett found himself revising his General Instructions for the Conduct of Naval Forces, which had been drawn up for an assault by night or at dawn on lightly defended beaches; new Instructions were now urgently required for a daylight assault on strongly held beaches.[40] So it took almost a year before it dawned on one of the leading defenders of the value of Jubilee's Lessons that the invasion assault would be qualitatively, not just quantitatively, different from even the largest raid.

The Navy, in fact, had fallen behind both the Army and the RAF in invasion preparations by May of 1943, despite the fact that one of the most immediate and tangible results of Jubilee was the creation of Force J, a naval assault force on brigade group scale for both training and operational purposes. Force J landed 3 Canadian division on D-Day and was commanded by Hughes-Hallett until April 1943. The files reveal, however, that Force J got off to a false start, suffering from an acute shortage of landing craft throughout this period, so that far less training than expected took place and no operations at all. Not that Hughes-Hallett, supported by Mountbatten, did not show his usual appetite for action. He always believed that a battleship could have turned the tide at Dieppe and now proposed the allocation of an R-class battleship for bombardment purposes or to act as a lure to provoke an air battle. Another possible use for such a ship was as a floating battering-ram to force a boom and land assault troops inside a defended port – one more reason for the strained relations between COHQ and the Admiralty. Admiral Little, C-in-C Portsmouth, was not enthusiastic about using the south coast as a training area and considered the Yukon exercises in June to have been lucky to avoid an E-boat attack; he was also against having such a 'conspicuous and valuable ship' in his Command. In any case, an Admiralty committee disagreed with Hughes-Hallett's reading of what a battleship might have accomplished at Dieppe, pointing out that broadsides fired from 10,000 yards could not have provided the accurate fire support at close range that was needed. In January, Hughes-Hallett was disappointed by the failure of landing craft crews 'to stay the course' while living under active service conditions in Portland harbour; by February, he could no longer justify maintaining his collection of ships and craft (without a battleship, needless to say) on an operational basis; by the end of April, Force J had lost most of its remaining craft to Husky and he asked to be relieved of his command, recommending the disbandment of what was left of Force J should there be no major operations in prospect before 1944.[41]

The implications of Bucknall's contribution to Rattle and the subsequent Corps Report on Study Week soon worked themselves out, helped by the rejection by the conference of the CDL tank for Neptune and despite misgivings at COHQ about mixed waves of assault craft. The War Diary of 3 Canadian division decreed on 25 August that future training and exercises in assault landings should envision an H-Hour between 'civil twilight [dawn] plus 2 hours and civil twilight plus 6 hours'.[42] It remained to decide on the most suitable mix of DDs, AVREs and specialized landing craft to match the beach defences and then to formulate the 21 Army Group doctrine for 'Assault on Defended Beaches'. The beauty of this decision was that its appropriateness was confirmed by technical developments between then and D-Day. What if, for example, a night assault had been settled upon, using CDL tanks and craft to dazzle the defenders, only for the Germans to compel a late change to daylight by introducing underwater obstacles with attached mines, as they did early in 1944? At the same time, however, it should be stressed that the likely availability of specialized armour was not in itself decisive but only one of a number of factors determining the selection of an assault in daylight.

To sum up: Jubilee was a hybrid of an operation. It marked the culmination of the policy of raiding which COHQ had been following since 1940. Attention at Richmond Terrace had necessarily (for want of resources and opportunities) been focused on hit-and-run raids using speed and surprise. Much of the equipment used at Dieppe, such as the LCA and R-craft, was designed with such operations in mind. Although COHQ had been set up to prepare the way for re-entering the Continent, nothing remotely resembling Neptune had been envisaged in 1940. It was still not envisaged in the summer of 1942 but there were signs that COHQ and other authorities had begun to experiment in adapting tactics to a strategic concept that included an assault in strength against a heavily defended coastline. The first use of LCTs at Dieppe, badly bungled as it was, was significant in this respect. In keeping with Jubilee's twofold nature, the flank landings relied on surprise but not the 'frontal' assault at 0520, which was denied the fire support that alone could have promised success. The Normandy invasion was planned on the assumption that tactical surprise would have been lost but in such a way that it could still be exploited if preserved.

The raid has received more than its share of attention, as already pointed out, because of its tragic outcome and the Canadian dimension. It is not completely fanciful to see in this phenomenon the historiographical reflection of that 'obtuse enthusiasm' for action for which Mountbatten and Hughes-Hallett have often been criticized.[43] For Mountbatten, what mattered most were operations, and he repeatedly allowed himself to be drawn into magniloquent but conjectural statements about saving ten lives on D-Day for every one lost at Dieppe – almost as though he would have exchanged his considerable achievements as CCO for a successful Jubilee. One consequence of this defiance has been the comparative neglect of those achievements, such as the Rattle conference, a neglect made worse by the failure to publish an official history of Combined Operations. There would have been a cross-Channel invasion in 1944 without Jubilee; whether there would have been one without Rattle, in many ways, according to Sir Harold Wernher, the supreme achievement of COHQ, is less certain.[44] Not nearly enough is known either about the history of amphibious warfare from an Allied, as opposed to a strictly British or American, point of view. For example, while British assault technique and equipment were designed for raiding, the Landing Ship Tank (LST), an indispensable component of all large-scale landings from Husky on, was a British specification. The LST did not belong on a raid or even in the initial assault of an invasion, but its significance in the Second World War has been compared with that of the tank itself in the First. Moreover, the technical representatives from COHQ who arrived in Washington at the end of 1941 had to plead with the US Navy to build the first 200 LSTs on British account.[45] The British were already thinking beyond a ship-to-shore fleet of small landing-craft to vessels which would be capable of landing tanks and MT on a shallow beach after a shore-to-shore ocean voyage. Accordingly, there was more to the story of Overlord than a scramble on the part of the British to adapt their tactics to the concept of a power-drive across the Channel – what Weigley termed 'pre-eminently an American military design'. The inspiration may well have been American but almost all of the nuts and bolts were of British invention.[46]

Finally, it may be that emphasis on only two aspects of Jubilee's context – radar and deception – has had a distorting effect on the interpretation of the raid's overall significance for the invasion. If so, each of the case studies at least serves to underline an important general point. First, *post hoc ergo propter hoc* reasoning is no substitute for historical analysis, particularly when dealing with such a complex technical subject as radar development. This particular leap of faith has characterized the justification of other Jubilee Lessons. It is beyond belief, for instance, that the invasion would have been launched by stealth, without a preliminary bombardment, save for Jubilee. Jubilee and Overlord should not be viewed as isolated events connected by a few direct strands but instead as belonging to a richly interwoven fabric, part of which stretches back to the work done by the Inter-Service Training and Development Centre before the war. Secondly, Jubilee lies on the far side of the great divide in the course of the war formed by the winter of 1942–43. This was too wide a gap for the raid to have contributed much if anything to strategic deception in 1944. It is impossible to read the Rattle documents or staff studies produced by COSSAC's staff in the early summer of 1943 without getting the impression of a fresh start, even a fresh – perhaps unwarranted – confidence. Before then, according to Morgan's senior Naval Intelligence officer, 'the attitude had to some extent been that the invasion would never take place, that the difficulties were insurmountable, that previous planners had been quite right in forming the most pessimistic opinions of its possible success, and that we should merely produce some plan with so many objections to it that nobody could ever accept it.'[47] There were two reasons for this change of attitude from gradually wearing the enemy out towards delivering a knockout blow: a growing awareness of the material and manpower resources of the United States; and above all the choice of a definite assault area and date.

NOTES

1. RG 24, C 3, 13,760, War Diary, G Branch, HQ 3 Cdn Inf Div, Dec. 1941–Feb. 1943: Notes on Address by Army Commander, 22 Dec. 1941.
2. Crerar quoted in *The Times*, 5 Aug. 1944; Mountbatten to Churchill Mann, 6 Oct. 1973, B 61, Mountbatten Papers. Wallace Reyburn, *Rehearsal for Invasion, An Eyewitness Story of the Dieppe Raid* (London, 1943). MG 30 E 463, Correspondence resulting from letter by Wallace Reyburn in Sunday Telegraph, 27 Aug. 1967, Hughes-Hallett Papers.
3. CAB 98/22, The Dieppe Raid (Combined Report), p. 46, para 367.

DEFE 2/334, HQ First Canadian Army: Intelligence Report, 8-5-1 Ops/56-1-1 Int, 22 Sept. 1942.
4. Major-General (retd) Churchill Mann, 'On the Real Purpose of the Dieppe Raid', *Canadian Defence Quarterly*, Vol. 9, No. 1, 1979, 57. See also John P. Campbell, 'Dieppe, Deception and D-Day', *Canadian Defence Quarterly*, Vol. 9, No. 3, 1980, pp. 40–44.
5. Mountbatten, 'Operation Jubilee: The Place of the Dieppe Raid in History', *Journal of the Royal United Services Institute for Defence Studies*, Vol. 119, No. 1, March 1974, pp. 25–30. This is the text of Mountbatten's speech to the Canadian Dieppe Veterans and Prisoners of War Association on 28 Sept. 1973 in Toronto.
6. OKM, 1 Skl I op 1661/42, Gkdos Chefs, 5.9.42 (betrifft Westwallartige Verstärkung der Küstenverteidigung im Westraum), RM 7/226. GenStdH, Abt Fremde Heere West III, 'Einzelnachrichten des Ic-Dienstes West Nr 28', 18.2.44, T 78/450. General der Pioniere und Festungen b ObdH, L III, 'Niederschrift über die Besprechung beim Führer über den Atlanktik Wall am 13 August 1942 (21.40–00.50 Uhr)', 14.8.42, T 78 317.
7. Der Chef des Generalstabes, Ob West, Heeresgruppe D, 'Entwicklungsmöglichkeiten der Lage im Westen', Ia Nr 55/42, Gkdos Chefs, 18.4.42, T 78 317; GenStdH, Abt Fremde Heere West, 'Notiz zur Feindlage', Nr 2203/44 gKdos, 12.4.44, T 78 451.
8. Hinsley, *British Intelligence*, Vol. 3 (2), pp. 60–1. ADM 223/287, 'German Views of Normandy Landing: Special Tactical Studies No. 30', MIRS/Lu/STS/30/44, 28 Nov. 1944.
9. CAB 80/72, COS(43)416(O), Operation 'Overlord', Report with Appendices, 30 July 1943: Appendix K, Annexure V, 'Enemy RDF Cover on the Western Front'.
10. AIR 20/1677, ADI(Sc), Air Scientific Interim Report, 'Hoardings', 21 Nov. 1942. Nissen, 'Dieppe in Retrospect', *Globe and Mail*, 19 Aug. 1982. AIR 8/895, Portal to Mountbatten, 2 July 1942, expressed annoyance at the way in which COHQ had gone about the experiments: 'I have just learned indirectly and quite casually that your people have been experimenting with "Gee" . . .' GEE was referred to as 'a well tried method' in one of the papers at the Rattle conference in June 1943 – RG 331, SHAEF, SGS, File 337/16: CO(R)3, 'Navigational Aids'.
11. Nissen, *Winning the Radar War*, p. 171; 'Dieppe in retrospect: it paid off on D Day', *Globe and Mail*, 19 Aug. 1972; letter to *Daily Telegraph*, 20 Sept. 1989.

12. AIR 16/733, 'Report on Operational Test Flight of Nine Defiant Aircraft fitted with Moonshine Equipment, 6 August 1942', 11G/S 600/2, 9 Aug. 1942. Martin Streetly, *Confound and Destroy: 100 Group and the Bomber Support Campaign* (London, 1978), p. 18.
13. AIR 20/1454, Air Commodore E.B. Addison to C. Paterson, GEC Wembley, 28 Aug. 1942; Minutes of Meeting held at HQ Bomber Command, 6 Oct. 1942, to consider radio counter-measures. AIR 14/277, Air Vice-Marshal Saundby to Groups, BC/S 28389/Sigs, 2 Nov. 1942.
14. AIR 20/1665, ADI(Sc), 'Air Scientific Report No. 111: Mandrel Results', 29 March 1943. AIR 20/1456, Addison to R. Cockburn (TRE), 18 May 1943; ECM Board Minutes, 27 July, 9 Sept., 28 Nov. 1943. AIR 20/1456, Air Vice-Marshal Saundby to Undersecretary of Air, BC/S.28387/Sigs, 8 Aug. 1942.
15. AIR 41/24, RAF Narrative, 'The Liberation of North West Europe', III, p. 31. AIR 40/2239, "Y" Intelligence for Combined Operations', including the report of 30 August by A.W. Bonsall (GC&CS). J.R. Robinson, 'Radar Intelligence and the Dieppe Raid', *Canadian Defence Quarterly*, Vol. 20, No. 5, 1991, pp. 37–43. Tait quoted in Toronto *Daily Star*, 2 Oct. 1967; Nissenthall, 'Dieppe Raid won key radar information', *Daily Telegraph*, 20 Sept. 1989. AIR 20/1456, ADI(K) Report No. 340/1943, 18 Aug. 1943: 'Some Recent German RDF Developments'.
16. DEFE 2/67, Operation Aflame, Leigh-Mallory to Mountbatten, 22 Aug. 1942; Minutes of Meeting at COHQ, 17 Sept. 1942. Air 16/763, Leigh-Mallory to Sholto Douglas, 11G/S 507/Ops, 28 Sept. 1942.
17. WO/4223, Brig. Loewen (GHQ Home Forces) to Brig. Curtis (War Office), HF/00/136/G(Plans), 31 March 1943. CAB 79/23, COS(42) 248th Meeting, 26 Aug. 1942; 264th Meeting, 16 Sept. 1942. DEFE 2/329, Williams, 'Notes for Press Conference, 1730, 20 August'. The COSSAC directive is to be found in CAB 80/69, COS(43)215(0), 25 April 1943. John P. Campbell, 'Operation Starkey 1943: "A Piece of Harmless Playacting"?' *Intelligence and National Security*, Vol. 2, No. 1, July 1987, pp. 92–113.
18. DEFE 2/538, Paper by Lt Cdr de Costobadie, 2 Dec. 1942.
19. DEFE 2/611, Operation Torrent, Cover for Overlord, COSSAC (43)39(First Draft for PSOs), 20 Sept. 1943. GenStdH, Abt Fremde Heere West III, 'Einzelnachrichten des Ic-Dienstes West Nr 28:

Bisherige Erkentnisse über das brit/amerik Kampfverfahren beim Landungsvernehmen von Nettuno am 22.1.1944', Nr 1763/44, 18.2.44, T 311/9. Masterman, *Double-Cross System*, p. 158.

20. Mountbatten made the comment about beach maintenance in the Summary of the Outline Plan for Rutter (P. 126, 11 May 1942) submitted to the COS – RG 24, G 3, 10,872, Files, HQ 2 Cdn Inf Div: 232C2(D 33).

21. Crerar quoted in *The Times*, 5 Aug. 1944. Hughes-Hallett, 'Memoirs', p. 304. Morgan, *Overture to Overlord*, pp. 155–7.

22. WO 219/1866, G-2 Records SHAEF, contains useful bulletins etc on lessons learned on Mediterranean operations. Roskill, *War at Sea*, III(1), pp. 183–4.

23. RG 331, SHAEF, SGS, Box 56, File 337/14, COSSAC Staff Conferences: Opening Address by Lt Gen F.E. Morgan, 17 April 1943. WO 106/4222, 'The Selection of Assault Areas in a Major Operation in North West Europe', CC(42)108, 5 Feb. 1943.

24. CAB 106/3, 'History of the Combined Operations Organisation, 1940–1945', p. 41. For the Mulberries, see CAB 106/969, 'History of COSSAC', p. 32; Hughes-Hallett claimed that the idea of artificial harbours came to him while listening to the anthem in Westminster Abbey in June 1943 (Ziegler, *Mountbatten*, p. 208).

25. CAB 106/1027, Draft Comments on PM's Memorandum on Roundup, HF 00/531/G(Plans), 25 June 1942; Notes for C-in-C [Home Forces] on points for discussion with the PM at 1730 hrs, 27 May 1942.

26. Hughes-Hallett, 'Memoirs', pp. 153–4. Stacey, *Six Years of War*, p. 329 footnote. PREM 3/256, Ismay to Churchill, 29 Dec. 1942. DEFE 2/561, Sledgehammer East. Wildman-Lushington to First Sea Lord [Mountbatten], draft letter, Feb. 1959, B 73, Mountbatten Papers.

27. DEFE 2/63, Operation Abercrombie, a Narrative; DEFE 2/65, Operation Bristle; WO 219/1933, Martian Report Nos 1 (2 June 1942), 2 (9 June 1942), 3 (16 June 1942), 12 (19 Aug. 1942).

28. 'Lagebeurteilungen durch Ob West vom 3.8.42 bis 18.10.43', T 78 311; WO 208/4308, 'German Defensive Strength in the Low Countries and North and West France', MI 14h/2/28/43, 8 March 1943.

29. CAB 80/72, COS(43)416(O): Operation 'Overlord', Report and Appreciation with Appendices, 30 July 1943, Appendices B,C,D and F.

DIEPPE AND D-DAY

30. DEFE 2/911, Meetings of COHQ (Executive) 1943: At the meeting on 11 November, Mountbatten decided that Jubilee's Lessons should be printed separately (Part B) and given wider distribution than the Report as a whole. Extract from letter, Mountbatten to Captain Knox (COHQ Liaison, Washington), 18 Dec. 1942, Box 1, Papers of Admiral H. Kent Hewitt, USN, Navy Yard, Washington DC.
31. RG 331, SHAEF, SGS, File 337/16: CO(R) 25, 'Rattle', Record of a Conference held at HMS *Warren* from 28th June to 2nd July to study the Combined Operations problems of Overlord, July 1943. Hughes-Hallett, 'Memoirs', p. 248. RG 24, C 3, 13,762, War Diary, G Branch, 3 Cdn Inf Div, Oct. 1943–May 1944: Oct. 1943 [for Pirate].
32. CAB 98/22, The Dieppe Raid (Combined Report), Part V, pp. 37–50.
33. ADM 199/1079, Hughes-Hallett, Enclosure No. 1 in 'Report of Proceedings', NFJ 0221/92, 30 Aug. 1942. DEFE 2/579, Report No. 109 by Historical Officer, 17 Dec. 1943. 'ENGLAND: Das Zusammenwirken der 3 britischen Wehrmachtteile . . . etc', Anlage zu 3 Abtlg Skl B Nr FL 18440/42 g, 8.10.42, T 1022 2363. DEFE 2/235, Operation Hadrian, CC(42)99(FSOs Draft). WO 232/36, 'Tank Policy', Director of the Royal Armoured Corps, 26 Feb. 1943.
34. RG 331, SHAEF, SGS, File 337/16: CO(R)25, 'Rattle' etc, pp. 16, 6; CO(R)11, 'The Assault, Pros and Cons of an Assault in Daylight or in Darkness', Memorandum by C.O.H.Q., 25 June 1943, pp. 77–9.
35. RG 331, SHAEF, SGS, File 337/16: CO(R)25, 'Rattle' etc, p. 16. Max Hastings, *Overlord, D-Day and the Battle for Normandy* (London, 1984), p. 315. Russell Weigley, *Eisenhower's Lieutenants: The Campaign of France and Germany, 1944–1945* (2 Vols, Bloomington, IN, 1974), I, pp. 51, 67. Ramsay quoted by Correlli Barnett, *Engage the Enemy More Closely, The Royal Navy in the Second World War* (London, 1991), pp. 796–7.
36. CAB 79/60, COS(43)51st Meeting(O), 22 March 1943.
37. ADM 223/287, Lt. Cdr. Richardson [SO(I), Neptune], 'Recollections of Planning Neptune'. Richardson commented that Hughes-Hallett 'whose brilliance of mind everyone immediately paid tribute to, was nevertheless a rather difficult master and an uneasy one. Actually his brain moved so quickly that conferences

with him rapidly took on the form of monologues on his part and this, though it is certain he never intended it to be so, had a repressive effect on many of his staff, and, if rumour be true, on large portions of the Army and Air Force.' Richardson admits to having been taken by surprise by the appearance of the underwater obstacles, although repeatedly warned about their possible appearance by Mountbatten. Hastings, *Overlord*, p. 315. CBC TV documentary, 9 Sept. 1962: Hughes-Hallett made the claim that Dieppe confirmed the Germans in a fatal policy of holding the invasion on the beaches; Crerar annotated this as an especially perceptive observation in the transcript of the programme in the Crerar Papers (MG 30 E 157, Vol. 26, National Archives of Canada).
38. DEFE 2/334, Report by Lt Col G.C. Reeves, GR/MM, 28 Aug. 1942; Report by Major B. Sucharov RCE, 55/OPS/41, SD(W), Appendix II, 2 Sept. 1942. Stacey, *Six Years of War*, p. 404, footnote. WO 106/4217, 'Development of Special Equipment for the Assault', WDC/P(43)18, 16 March 1943. WO 165/132, Half-yearly Report on the Progress of the Royal Armoured Corps, 1 July–31 Dec. 1942, Appendix U.
39. Captain B.H. Liddell Hart, *The Tanks* (2 Vols, London, 1959), II, p. 321. WO 165/133, Half-yearly Report of the Progress of the Royal Armoured Corps, 1 Jan.–30 June 1943, Minutes of 27th Meeting on 30 June.
40. WO 166/1037, War Diary, G Branch, 1 Corps HQ, 1943: Report of Study Period, 23–26 June (Appendix II to WD for July). RG 331, SHAEF, SGS, File 337/16: CO(R)25, 'Rattle' etc, pp. 14–15.
41. ADM 179/254, Establishment of a Raiding Force on the South Coast: Hughes-Hallett to C-in-C Portsmouth, J0100/6, 3 Nov. 1942; Hughes-Hallett to C-in-C Portsmouth, J 0100/37, 26 April 1943; Hughes-Hallett to Mountbatten, J0100/37, 25 Feb. 1943; comment by Admiral Little on Staff Minute Sheet, 21 Aug. 1942. ADM 199/1199, 'Reports on Various Combined Operations 1942', R.A. McGrigor's Committee Report, 5 Dec. 1942. DEFE 2/727, Memo No. J 0208/31, 30 Jan. 1943; DEFE 2/458, Hughes-Hallett to Mountbatten, J0513/1, 18 March 1943.
42. DEFE 2/1017, Meetings of CO Executive 1943, 74th Meeting, 19 July 1943. RG 24, C 3, 13,761, War Diary, G Branch, HQ 3 Cdn Inf Div, March–Sept. 1943: 3CD-4-3-4, 'Timing of Assault Landings', 25 Aug. 1943. When Hughes-Hallett was amending

and amplifying his superseded General Instructions, he took into account a Memorandum by Crerar which indicates that the Canadian Army had also been working on the problems of a daylight assault on a heavily defended beach; Crerar stressed the need to gap beach minefields and wire *before* the infantry touched down but was still attached to the 'Jubilee' idea of mobile forts (ADM 179/306, 'Training for Daylight Assault', Memo No. 857 [Hughes-Hallett], 10 Sept. 1943).
43. Hamilton, *Monty*, I, p. 546.
44. Wernher cited by Ziegler, *Mountbatten*, p. 214. Ziegler himself describes the Rattle conference as 'a masterpiece of presentation', but goes too far in suggesting that Overlord was essentially the plan that had been in gestation at COHQ for the previous year, except in the sense that Normandy had always been Mountbatten's choice for the assault area.
45. There is interesting correspondence between Captain T. Hussey and Mountbatten dating from 1970 and 1974 on the subject of the LST (B 49, Mountbatten Papers). The same file contains comments by Hussey and Laycock (1960/61) on the shortcomings of Roskill's Official History where Combined Operations were concerned.
46. Weigley, *Eisenhower's Lieutenants*, I, p. 46. CAB 106/7, 'Summary of history and function of Combined Operations Organization'.
47. ADM 223/287, Lt Cdr Richardson, 'Recollections of the Planning of Neptune'. Although still listed in the Class List without an asterisk, this piece now (1992) seems to have been withdrawn from circulation at the PRO.

Glossary of Codenames for Operations

Abercrombie	Hardelot	21/22 April 1942
Aflame	Berck	October 1942
Aimwell	Alderney	August 1942
Barricade	Barfleur/St Vaast	14/15 August 1942
Biting	Bruneval	27/28 February 1942
Blazing	Alderney	May 1942
Bristle	Etaples	3/4 June 1942
Cauldron	Varengeville	19 August 1942
Chariot	St Nazaire	27/28 March 1942
Dryad	Isle Casquets	2 September 1942
Foxrock	St Valery's Somme	1/2 June 1942
Husky	Sicily	10 July 1943
Hadrian	Cherbourg Peninsula	1943
Imperator	Bresle/Somme	July 1942
Lancing	Boulogne/Le Touquet	May 1942
Myrmidon	R Ardour defences	4/5 April 1942
Neptune	Ouistreham/Varreville	6 June 1944
National	St Jean de Luz	June 1942
Sledgehammer	Cherbourg Peninsula	1942/43
Starkey	Pas de Calais area	9 September 1943

Index

Abbeville, proposed bombing of, 136, 137
Abercrombie (Operation), 6, 106, 107, 214
Abwehr (German Intelligence), 20, 88–9, 96; and agents, 23, 56, 176; reports of invasion threats, 162–3; role of, 42–3
Abwehr III Counter-Intelligence, 22–3
Abwehr stations, 41, 47
Admiralty: authority of, 167; Naval Intelligence (NID), 52, 170; relations with COHQ, 224
ADRDE (Air Defence Research and Development Establishment) Malvern, 147
aerial reconnaissance (GAF), 43, 98, 100–1, 161; and use of radar, 132–3
aerial reconnaissance (RAF), to confirm Ultra, 171
Aflame (Operation), 206–7
agents: Abwehr agents in England, 22–23, 53, 91; German use of, 41–2, 43, 176; inventing, 44–5, 54–6; sub-agents, 54
agents, double (XX), 22, 44, 46, 48, 59–60; records of reports from, 31–2, 60
agents, individual: A 1392 'Princess', 48–9; A 3924, 41; Bertrand, 25–6; Dragonfly, 22, 46, 48, 51; the Druid, 49–50; Garbo, 49; Mutt and Jeff, 22; Ostro, 54–6; R 3764, 42, 47; Tate, 22, 47–8
AI 3b *see* Air Intelligence
Aimwell (Operation), 65–6
Air Force, German *see* GAF (German Air Force)
Air Force, Royal *see* RAF (Royal Air Force)
Air Intelligence, AI 3b, 187, 188–9
aircraft: Blenheims, 12; Bostons, 12; FW 190, 186; Hurricanes, 9, 11; Junkers (Ju) 86P, high altitude bomber recce, 100–1; Spitfire IX, 101, 186; Spitfire V, 186; Typhoon, 186
aircraft losses at Dieppe, 187–9
alarm rockets, naval signal station, 142
Albrighton (destroyer), 104, 128, 133, 135
Alderney, 162, 164, 172, 221; Operation Aimwell, 65–6; Operation Blazing, 65, 140, 164; Operation Dryad, 53; Sledgehammer Central, 67
anti-tank defences, 218
Antwerp, 96, 215
armour *see* tanks
Army (British) 1 Corps, 222
Army (British): 21 Army Group, 225; 79 Armoured Division, 222; coastal defence system, 134; wireless security, 180–81
Army (Canadian), 2, 9, 182; 14 Tank Battalion, 9–10; 2 Infantry Division, 3, 197; 3 Division, 197, 219, 224–5; Exercise Yukon II, 70; Field Security, 5; Royal Regiment of Canada, 10–11
Army (German), 113, 134, 221; 15 Army, 50, 110
Army (German), *see also* corps; divisions
assaults, 217; on defended beaches, 225; frontal, 217–18; 'the problem of', 220

INDEX

Atkin, Ronald, *Dieppe 1942*, 18
Atlantic Wall, Hitler's concept of, 82, 200
Auderville, Freya radar at, 138
Avalanche (Operation), Salerno, 210
AVRE (Armoured Vehicle Royal Engineers), 222

B Bericht, intelligence reports, 179
B-Dienst: Brughes station, 174–5; interceptions, 106, 126, 171, 176; role of, 24, 177, 179
Backchat (Operation), 59
Baedeker raids, 101
Baillie-Grohman, Rear-Admiral H.T., 8, 26, 27, 108, 136
Barbarossa (Operation), 104
Barfleur, Operation Barricade, 65
barges, 61, 100
Barker, General R., 220, 221
Barricade (Operation), 58–9, 65, 122
battleship, request for refused, 8, 224
Bayonne, Myrmidon raid, 65, 66
BBC, broadcast to French coast, 69
beaches: Berneval, 7, 8; Overlord appreciation of, 215; Pourville (Green), 8, 10, 12; Puits (Blue), 8, 10–11, 219; Red (town), 8, 9; Varengeville, 8; White (town), 8, 9
Beesly, Patrick, 159, 161, 170–1, 173
Bennett, Ralph, 159, 160
Berck, radar at, 143
Bernal, Professor, 148
Berneval, 124, 142; beach, 7, 8
Bevan, Colonel J., 62, 89
Birmingham, Baedeker raids on, 113
'Bismarck' headland, 10, 11, 12
Blazing (Operation), 65, 140, 164
Bletchley *see* GC & CS
Bock, Field Marshal von, 84
bombing: of coastal batteries, 221; diversionary, 127, 140–1; as part of raid plans, 8–9, 69, 136–7, 140–1
bombing raids, Baedeker, 101
Bonatz, Heinz, 179
Boniface, and Ultra, 159

Boulogne, 26, 93, 125; operations against, 6, 66, 67
Bramesfeld, Captain, 93, 112
Brest, 94
bridgehead, seizure of *see* Sledgehammer
Bridport, 102
Bristle (Operation), 66, 107, 214; radar and, 122, 128–9
Bristol Channel, 100
British Communist Party, 28
British intelligence, official history authorized, 24–5, 159–60
British intelligence, *see also* agents; Operational Intelligence Centre (OIC); radar; Special Intelligence; Ultra; Y intelligence
Brittany, speculation about landings in, 91, 92
Brocklesby (destroyer), 109, 148
Bruce Lockhart, Robert, 6
Bruneval raid, 46, 65, 171; and GAF radar, 122, 133, 138–9
BST (British Summer Time), 123–4
Buchan, Alastair, 9
Buchheit, Gert, ex-Abwehr, 23
Bucknall, Lieutenant-General G.C., 223, 224
Butler, J.R.M., 33

Calais, 93, 105
Calpe (destroyer), 9; signals to, 146, 147, 148, 171, 179
Calvo, Spanish journalist, 49
Calvocoressi, Peter, *Top Secret Ultra*, 159–60, 188
camouflage, lack of on British shipping, 79
Campbeltown (destroyer), 127, 128
Canadian army *see* Army (Canadian)
Canaris, Admiral Wilhelm, Chief of Abwehr, 20, 42
Cap d'Alprech, radar at, 143
Carls, Grand Admiral, 60
Casa Maury, Wing Commander the Marquis of (COHQ), 27, 163–4, 165–6

236

INDEX

Casablanca, 42; conference, 207, 216
Cave Brown, Anthony, 58–9, 69, 149, 160, 199
Cavendish (Operation), 55
caves at Dieppe, 25, 151
CD/CHL radar *see* radar, CD/CHL
Channel Dash, 126, 148, 173
Channel Islands, 103, *see also* Alderney
Chariot (Operation) *see* St Nazaire
Cherbourg: defences, 67, 93, 215; Operation Gabriel, 66
Cherbourg peninsula, Operation Barricade, 58–9
Chichester, Assault Landing Craft at, 78
Churchill, Winston, 64, 68, 211; on effect of small raids, 59, 69
ciphers *see* codes
coastal batteries, 61–2, 108, 134, 221
coastal defences (German), 22, 85, 218; improvement of, 200, 214, 220–1
Cockroft, J.D., 147–8
codenames, procedure on, 53
codes: in British army R/T, 181–2; British naval ciphers broken, 14, 24, 179; cipher security, 174; Enigma (Cockroach), 166; GAF Red key, 161; ZTPG (Heimisch) key, 30, 161, 167
codes, *see also* Ultra
COHQ, 5, 130, 150–1, 221; and cancellation of Rutter, 25, 26–7; Examination Committee, 27, 51, 65, 66–7; 'Pros and Cons of Assault in Daylight or Darkness' (paper), 219–20; rift with ISSB over Jubilee, 52–3
collision *see* Jubilee (Operation), collision with convoy
Combined Commanders: Overlord, 215–16; pressures on, 64, 66
Combined Operations: base on Isle of Wight, 62, 149; History of, 33, 212; stage management of, 196

Combined Operations, *see also* COHQ
Combined Operations, Chief of, duties of, 33
Combined Operations, Chief of, *see also* Mountbatten, Lord Louis
Communists, in Dieppe, 180
convoys: British, 54, 87–9, 109; German, 103, 112; PQ, 16, 18, 113
convoys, *see also* Channel Dash; Jubilee (Operation), collision with convoy
corps (German): LXXXI, 45, 59; LXXXIV, 59, 184; SS Panzer, 113; Waffen-SS Panzer, 80
COS (Chiefs of Staff): pressures on, 66; and Rutter security, 4; value of deception planning, 62; view of Hitler, 61
COSSAC *see* Morgan, Lieutenant-General F.E.
Coughdrop (Operation), 164
Coventry, air raid, 24, 25
cover: diversionary, 208; need for defensive, 197–8
cover, *see also* strategic deception
cover plans: definition of, 56–7; Jubilee and, 28, 70, 92, 197–8; Rutter, 28, and security, 57–8; 'special means', 57–8, 70; for Torch (Overthrow and Solo One), 46; use of cancelled operations, 51, 67
Crerar, General H.D.G., 70, 146, 196–7, 209–10
Cripps, Sir Stafford, 151

D-Day, xiv–xv, 215–16, 221; radar counter-measures, 202, 204
D-Day, *see also* Neptune; Overlord
de Costobadie, Lieutenant-Commander A.N.P., 207
Dearson, K., 146
deception: formal, in Allied strategy, 60–1; Jubilee and, 70–1, 197–201, 226; and presuppositions, 90–2

INDEX

deception planning: and cancelled operations, 51, 67; long-term, 62; W Committee, 27
decrypts: release of, 30–1, 159, 165; time taken for, 167, 171, 172; ZTPG series, 30
decrypts, *see also* Ultra
Delight (destroyer), 138
Denmark, 80, 162
Denovan, Lieutenant J.J., 222
destroyers, 7, 109, 127, 128, 138, 148; *Albrighton*, 104, 128, 133, 135
Detailed Military Plan (JG One), 7, 184; captured at Dieppe, 98
Devers, Lieutenant-General J., 216
Dewdale (Landing Ship (Gantry), 97
Dieppe, xiii; bombing threatened (June 1942), 69; choice of, 213; defences strengthened, 214–15; disadvantages of, 93, 186; German map exercise, 18–19; low priority of, 102–3, 113, 130; radar stations at, 122–23, 132–3; repeat raid predicted, 56
Dieppe, *see also* Jubilee (Operation)
Dietrich, Otto, 82
diplomatic sources of intelligence, 42, 43, 51, 68, 86
disinformation: agent Ostro, 54–6; Operation Hardboiled, 44–5
divisions (German): 2 Panzer, 184; 6 Panzer, 80; 7 Panzer, 80; 10 Panzer, 25–7, 50, 80, 184, 185; 23 Infanterie, 80; 24 Panzer, 17; 106 Infanterie (ID), 21; 110 Infanterie (ID), 20, 21, 182–3; 302 Infanterie (ID), 3, 45, 50; (and Jubilee), 59, 141, 142; (whereabouts uncertain), 20–1, 182–3, 184; 321 Infanterie (ID), 111, 112; 712 Infanterie (ID), 184; Das Reich SS motorized, 80; Defence, 104, 112, 129–30, 134; Grossdeutschland motorized, 16, 80, 84, 183; SS Adolf Hitler, 50, 80, 83; SS Totenkopf, 80
divisions (German), *see also* Army (British); Army (Canadian); Army (German); corps; GAF
documents: captured at Dieppe, 96–8, 109, 174, 198–9; captured at St Nazaire, 175
documents, *see also* records
double agents *see* agents, double
Douglas, Air Chief Marshal Sholto, 186
Dryad (Operation), 53
Dunfermline, Mountbatten 'embarked' at, 44
Dunkirk, disadvantages of, 93

E-boats, 103, 105, 150, 172; flotillas, 108–10; radar interception of, 177; radar search receivers on, 107; risk of ambush by, 108
early warning system (German coastal), 107–8, *see also* radar
El Alamein, 218
Enigma, 24, 31, 49, 166, 175–6, 188, *see also* Ultra
Etzdorf, von, Foreign Office representative at OKH, 110–11
exercises *see* Pirate; Yukon

Falmouth, 78
Farago, Ladislas, *The Game of the Foxes*, 48–9, 51
Fécamp, 93, 104; Seetakt radar at, 129–30, 131
Fernie (destroyer), 9, 166
Fidelitas (ore ship), 112
fire support, 8–9, 217–18, *see also* bombing
Fleming, Commander Ian, RNVR Naval Intelligence Division, 52, 166, 170
FMZ (air raid reporting centre), 138, 141, 142, 143, 167
Force J, naval assault force, 224

238

INDEX

Foreign Armies West: evaluation of intelligence, 46–7, 87, 89; subordinate to OKH, 88
Foreign Office, warnings to French coast, 69
Fortitude South (Operation), 198, 200–1, 208
Foxrock (Operation), 178
France: air penetration into, 185–6; expectation of Second Front, 69; invasion alarms (Abwehr report), 162–3
Franks, Norman, *The Greatest Air Battle*, 18
Freya *see* radar
Frölich, General, 14
Führerbefehle (Hitler's personal orders), 16–17

Gabriel (Operation), 66
GAF (German Air Force), 25; 3 Air Fleet, 14, 15, 78, 79, 100, 101, 102; 7 Flieger, 15, 16, 80, 184; III/KG 26 torpedo bombers, 18–19, 113, 188; III/KG 53 and III/KG 54, 188; IX Fliegerkorps bomber-recce (Ju 88), 101; Hermann Goering brigade, 15, 16, 80, 184; and Jubilee air battle, 178–9, 187, 205–8; limitations of, 89–90, 101, 102, 208; limited Ultra intelligence about, 161, 188–9; map exercise 15 Aug 1942, 18–19; R/T and W/T transmissions monitored, 177–8; radar *see* radar; reconnaissance, 78, 79, 91, 98, 107, 181, 214; warning and control system, 185–6; Y intelligence, 178
GAF (German Air Force), *see also* aircraft
Garth (destroyer), 11, 12, 148
GC & CS (Government Code and Cipher School), 24, 30, 48, 49, 159, 217, *see also* Ultra
Gehlen, Lieutenant Colonel, 43

German Defence Measures, Ultra information on, 161–2
German defensive system, effect of agents' reports on, 45, 46
German Intelligence: fragmented nature of, 86–7; Hitler's selective use of, 86–8; Naval, 87, 174–6, 179; and Ultra, 160, 161, 174
German Intelligence, *see also* Abwehr; agents; Enigma; photo reconnaissance; radar; RSHA (Amt VI); Y intelligence
German Naval Staff: interpretation of coastal activity, 90–1; and timing of British invasion, 94–5
German Naval Staff War Diary, agents' reports in, 23, 55–6
German Navy: coastal radar, 125; Defence Divisions, 104, 112, 134; Naval Group West, 93–4, 102, 105–6, 129–30, 174, 176; Naval Intelligence, 87, 174–6, 179; records held by ONI, 19–20
German Navy, *see also* B-Dienst
Germany: anti-invasion planning, 70, 200, 208; and progress of war in summer 1942, 81–2
Giskes, Hermann, Abwehr controller, 23
Gneisenau (battleship), 173
'Gneisenau' heavy railway battery, 102
Goebbels, Joseph, 68, 81–2, 83
Goering, Hermann, 15
Golden Sunbeam (drifter), sinking of, 179–80
Grängesberg (ore ship), 112, 176

Haase, Colonel-General Curt, 50, 111
Haase, Lieutenant-General Konrad, 3
Hadrian (Operation), 218
Haines, Lieutenant-Colonel, 164–5
Halder, Colonel-General Franz, 81, 84, 85, 87, 95
Hamilton, Nigel, *Monty*, 2

239

INDEX

harbour protection boats, at Dieppe, 108
Hardboiled (Operation), 44–5, 62, 162
Hastings, Max, *Overlord, D-Day and the Battle for Normandy*, 220, 221
Haydon, Major-General J.C, 19
Head, Colonel, 151
Hinsley, F.H.: *British Intelligence in the Second World War*, 24–5, 26, 171, 173; on Sigint, 182–3, 217; on Ultra, 160–1, 188
Hinsley, F.H. and C.A.G. Simkins, *Security and Counter-Intelligence*, 27–8
Hitler, Adolf, 1, 50, 165; and coastal convoys, 103, 104–5; expected invasion areas, 17, 200, 201; and German Intelligence, 46, 86–8, 175–6; importance of Norway, 87–8; reinforcement of French coast, 15, 79–80, 81; telegram, 14–16
Hobart, Major-General P.C.S., 1, 222
Hoffmann, Karl Otto, *Geschichte der Luftnachrichten Truppe*, 122, 140, 141, 143
Holland, 95, 110, 113
Howard, Michael, *Strategic Deception* (Official History of Intelligence), 27–8, 50
Hughes-Hallett, Captain John: on beach landings, 9, 10; on *Calpe*, 148, 171; chairman of Low Cover RDF Committee, 125; character, 231n; commanding Force J, 224; difficulties of Channel operations, 109, 137, 210, 219; dispute with Reyburn, 197; enthusiasm for Jubilee, 196, 226; planning of raids, 57, 65, 151, 212, 217; radar navigation aids, 148; radar and tactical surprise, 28, 29, 142; radar tracking of Jubilee, 140, 146; on tactical surprise, 28, 29, 142, 150
Hunt, Sir David, on Ultra, 159
Husky (Operation), 207, 209–10, 226

IFF (Identification Friend or Foe), for radar, 126
Illustrated London News, photographs of R-craft, 96
Iltis (German torpedo boat), 104, 126
Imperator (Operation), 66
India, 88
infantry, landing of, 218
intelligence, sources of, 174, 219
intelligence, *see also* British intelligence; German intelligence
Inter-Service Security Board (ISSB), 4, 19; rift with COHQ, 52–3
Inter-Service Training and Development Centre, 227
Interservice Codeword Index, 53
invasion: German expectations of, 81, 87, 94–5, 101, 162–3, 208; specialized nature of, 209, 211, 223
Irish Sea, 100
Irving, David: access to German records, 13–17, 19–20; agents' reports, 22, 23, 47; Jubilee and Ultra, 164
Isle of Wight, 5, 45, 79; Combined Operations base, 62, 149

Jacob, General A., 80
Japan, entry into war, 87, 88
Jodl, General Alfred, 42, 84, 87
Joint Intelligence Committee (JIC), 19, 49, 52–3, 183
Jones, R.V.: 'Interim Report on German Coast Watching RDF Stations', 130; *Most Secret War*, 124; radar expert, 122, 125, 137–8, 152, 203, 205
Jubilee (Operation): air battle, 186, 187–8, 206, 207; collision with convoy, 11, 28, 30,

240

107, 113, 123, 142, 148, 171–2; Combined Plan (JJ One), 6–7, 33; Combined Report (defensive cover), 197–8; (radar), 28, 124, 170; (security), 5, 6; deception lessons from, 197–201; detailed planning of, 6–8, 95; disastrous effect of mis-timings on, 10–13; diversionary bombing dropped from plans, 140–1; effect on GAF fighter force, 205–8; inadequacy of 'breach of security' hypothesis, 50; lessons from, 196, 207, 212, 227; lessons from and Rattle conference, 216–22, 223, 226; lessons misread by Germany, 199; nature of, 67, 68, 196, 225; numbers of landing craft, 100; Operation Orders, 6–7, 10–11, 28; (and LSIs), 147; (captured), 96–8, 109, 174, 198–9; Outline of Operation (Naval) JNO One, 7, 98; Overlord's debt to, 196–7, 211; radar lessons from, 201–5; radar tracking of, 139–41; reasons to cancel, 152–3; role of radar, 122–53; Ultra and, 161, 166, 167, 169; verdict on (author's), xiv; wireless silence broken, 179–80
Jubilee (Operation), *see also* Dieppe

Keegan, John, *The Face of Battle*, 27, 33
Kennard, Captain L., 180–1
Kessler, General Ulrich, 14, 15, 179
Knightley, Phillip, 160
Kootwijk wireless station, 23
Kuntzen, General, 185

Lageberichte West, 182, 184–5
Lancing (Operation), Boulogne, 66
landing craft, 64, 141, 219, 223, 224; German estimates of, 78, 95–8, 100, 102; LCA (Assault), 11, 78, 218; LCT (Tanks), 7, 9–10, 78, 96, 218, 225; LSIs, 5, 7, 97, 109
landings, daylight, 223, 225
landlines, German secure, 48, 183–4
Le Havre, 57, 93; defences, 67, 215
Le Tréport, 41, 169; radar at, 130, 131, 143, 145
Leasor, James, *Green Beach*, 29, 123, 141, 146
Leigh-Mallory, Air Vice-Marshal Trafford: and Jubilee, 1, 9, 140, 187, 205–6, 207; Operation Aflame, 206; plan to capture Boulogne, 66; and radar tracking, 132, 136, 137, 140; and Rutter, 136, 137, 186; on tactical surprise, 149–50
Lewin, Ronald, *Ultra Goes to War: The Secret Story*, 30
Liddell Hart, Basil, 70, 222
lights, coastal, 126
Lisbon, 42, 49
Liss, Lieutenant-Colonel Ulrich, 43, 87–9
Little, Admiral Sir Charles, 210, 224
Locust (monitor), 11–12
Loire estuary, radar on, 127
Lorient, Operation Coughdrop, 164
Lovat, Lord, 5
LSI (Landing Ship Infantry), vulnerability of, 5, 7, 97, 109
LST (Landing Ship Tank), significance of, 226
Luftwaffe *see* GAF (German Air Force)
Lyme Bay, 102, 110

McNabb, Brigadier, 151
McNaughton, General A.C.L., 70
Mandrel *see* radar counter-measures
Mann, General C. Churchill, 197, 198, 199
Marshall, General, 65
Martian Reports, 20, 183–4, 214; No 2 (9 June), 92; No 4 (23 June), 26; No 5 (30 June), 27, 140; No 7 (14 July), 113, 184; No

INDEX

12 (19 August), 92; No 13 (26 August), 183; No 14 (2 Sept), 101; No 17 (23 Sept), 184
Masterman, Sir John, 24, 57–8, 208; *The Double-Cross System in the War of 1939 to 1945*, 22, 48
Mediterranean, lessons from for Overlord, 209–10
Menzies, Sir Stewart, 28, 164
MGB (motor gun boats), 11, 104, 105
MI 5, 6, 22, 31, 49
MI 6 (Secret Intelligence Service), 25, 28, 31
Middle East, British reinforcement of, 87, 88
mine-barrier, 105–7
minefields, Worthing, 60, 61
minelaying, 104, 106, 129–30, 173–4; British, 93–4
mines: ground (inshore), 105–6; moored, 105, 106
minesweepers, as patrol boats, 107
minesweeping, for channel convoys, 103
Minesweeping Flotillas (German), 103, 107, 112
Molotov, V.M., visit to London, 78–9
Montagu, Ewen, 51, 59–60; *Beyond Top Secret Ultra*, 48
Montgomery, Field Marshal Lord, 26; 'Lessons Learnt', 196
Mordal, Jacques, *Dieppe*, 29, 139, 141
Morgan, Lieutenant-General F.E.: COSSAC, 70, 207, 210, 211, 216, 222–3; *Overture to Overlord*, 220
Mosley, Leonard, *The Druid*, 49–50, 51
Most Secret Source (MSS), 45, 159
Mountbatten, Lord Louis, 43–4
Mountbatten, Lord Louis, Chief of Combined Operations: and deception for Overlord, 199, 200; enthusiasm for Jubilee, 151, 196, 226; enthusiasm for raids, 58, 65, 151; on fire support, 217; 'Lessons' from Jubilee, 207, 216; and origins of Jubilee, 4, 33–4, 51; on strategic surprise, 149; and tactical surprise, 70
MTB (motor torpedo boats), 104, 105, 176
Mulberry harbours, 212
Murphy, Peter, 68
Mussolini, Benito, 82
Myrmidon (Operation), 65, 66, 98

National (Operation) (St Jean de Luz), 27
naval fire support, usefulness of, 210
Naval Intelligence (Admiralty) (NID), 52, 170
Naval Operation Orders, captured at Dieppe, 98, 174
Naval Staff Battle Summary, on radar warnings, 147
navigation, problems of, 7, 128, 140
navigation aids: GEE, 148; radar, 148
Navy *see* German Navy; Royal Navy
Neptune (Operation), 209, 225, *see also* D-Day; Overlord
Neville, Colonel R., 53
Next of Kin, film, 6, 45, 57
Niehaus, Werner, *Die Radarschlacht*, 123, 141
night fighter control (GCI), 138–9
night fighter zones, and radar, 203
Nissen(thall), Jack, 146, 204; and Freya radar at Dieppe, 135, 140, 201–2; *Winning the Radar War*, 29, 123, 124
North Channel, 100
Norway, 113; threat of invasion, 44, 46, 162, 164

Oculist, and Y intelligence, 138, 139, 145–6, 203, 205
Office of Naval Intelligence (ONI) (US), German naval records held by, 19–20
Official History, Ultra as missing factor in, 159

242

INDEX

Official History of Deception, use of XX agents, 59
Official Secrets Act, decrypts still held under, 31
OKH (German Army General Staff): on captured plans, 174; conflict of authority with OKW, 88; Foreign Armies branch, 46; on likelihood of British landing, 95; preoccupation with Russian front, 84
OKW (German High Command): control of operations in West, 88; on likelihood of Second Front, 84; War Diary, 32, 43–4, 46–7, 183
operation planning, need for flexibility in, 218–19
operational categories, limited strategic goals, 86
Operational Intelligence Centre (OIC) (Admiralty), 30, 160, 161, 167, 170–1, 173; air intelligence (AI 3b), 187, 188–9
operations *see* individual operation names
order of battle, intelligence about German, 172–4, 182–4
Oslo Report, on GAF radar, 137
Ostend, 65
Ostro (Paul Fidrmuc), agent, 54–6
Overlord (Operation): choice of beaches for, 211–12, 215; cover and deception for, 198; debt to Jubilee, 196–7, 201; and GAF fighter force, 205; planning of, 220–1; relationship to Jubilee, 208–10
Overlord (Operation), *see also* D-Day; Neptune
Overthrow (Operation), 46, 51, 55, 70, 199
Ozets *see* radar plotting centres

parachute troops, 97–8, 180
Paris, Operation Imperator, 66
Pas de Calais, beaches, 215
Pastorius (Operation), 42
patrol boats (Vorpostenboote), 93, 107, 108, 112
Peis, Günther, 47–8, 51
Philby, Kim, 50
photo reconnaissance (PR), 25, 171; GAF, 78, 100, 214; of radar stations, 131, 137–8; RAF, 214; of shipping, 95–6, 98
Pirate, exercise (October 1942), 216
placenames, German spelling of English, 60
Pointe d'Ailly, Seetakt radar at, 129–30, 131, 145, 152
Political Warfare Executive (PWE), 67–8, 69
Portland, 78, 139
ports, and invasion plans, 199–201, 212
Pourville, 142; Green beach, 8, 10, 12
POWs: after Jubilee, 13; GAF officer, 18–19; interrogation by OKH, 43, 95
PR *see* photo reconnaissance (PR)
Prince Albert (LSI), 97, 128–29
Princess Astrid (LSI), 11, 79
Princess Josephine Charlotte (LSI), 79
propaganda, 68, 82
public opinion, pressure for Second Front, 68, 69–70
Public Record Office (PRO), decrypt series in, 30–1
public records, access to, 17–18, 30–2
Public Records Act (1958), 15, 17, 31
Public Records Act (1967), 17–18
Puits, 141, 143; Blue beach, 8, 10–11, 219

Queen Emma (LSI), 11

R-boats, 104, 136, 150, 172, 175
radar, 28, 122, 140–1; anti-jamming devices, 135; ASV, 148;

243

INDEX

at Barfleur, 65; at Berck, 143; at Dieppe, 28–9; on British shipping, 104; British, superiority of, 174, 177; CD/CHL (Coastal Defence/Chain Home Low Flying), 134, 139; (jamming of), 127, 147; Freya (capabilities of), 131–2, 133, 135, 138, 139; (counter-measures), 202, 203, 204; (Plage St Cecily), 128, 129; (PR intelligence about), 137–8; Freya (F 28) Dieppe, 102, 122–3, 124, 201; (and Jubilee), 28, 140–1, 145, 149; GAF, 28, 29, 102, 131, *see also* radar, Freya; Würzburg; (eccentricity of), 135; GEE, 202; German, 128, 145–6; (intelligence about), 122, 124, 130–1, 137–8, 205; German Navy, 125; Giant Würzburg, 202; Hoarding (Mammut), 202; IFF devices, 126; jamming, 126–7, 129, 147; Lorenz equipment, 127; navigational, 104, 148; and night use, 131, 203; Operation Blazing, 140; Operation Bristle, 128–9; Operation Chariot, 127–8; risk of detection by, 140–1, 152–3, 201; role of in Jubilee, 145–7, 152–3, 226–7; Seetakt, 125, 127, 130; Seetakt coastal chain, 125–7, 129–31, 152, 201; Seetakt stations, 129–30, 131, 143, 145, 152; Type 271, 147; Type 272 Naval, 148; Type 286, 148; Type 290, 148; Type 78 D/F, 148; Würzburg, 132, 133, 135; (Bruneval), 122; (capabilities of), 138, 139; (Plage St Cecily), 128, 129

radar counter-measures: D-Day, 204; Mandrel 120 MHz, 202–4; Moonshine, 202–3, 205

radar plotting centres (Ozets), 125; Boulogne, 125, 142, 143, 145
radio transmission, 177
Raeder, Grand-Admiral Erich, 86, 101; inspection of defences, 91, 103, 112, 125
RAF (Royal Air Force), 103, 134, 209; and air superiority, 101, 185–6; Bomber Command, 203; and Dieppe air battle, 187–9; Fighter Command, 101, 186; and GAF radar, 138, 203; night operations by, 109–10; No 11 Fighter Group, 185–6; Wellington squadrons, 4; Y intelligence, 177–8
RAF (Royal Air Force), *see also* aircraft
Raider B *Komet*, sunk off Cherbourg, 171
raids: cancellations of, 64–5, 150–1; cross-Channel, 3, 63–5, 67–8, 225; different from invasion, 209, 223; intended effect on GAF capability, 206, 207; minor, effect of on German security, 58–9, 69; varying size of, 57; von Rundstedt's view of, 85
Ramsay, Admiral Sir Bertram, 210, 220
Rattle conference at Largs, 216, 219, 223, 226, 227
records: lost and destroyed, 20, 32, 87, 124–5; reliability of, 32–3
records, *see also* documents
Reeves, Lieutenant Colonel G.C., 222
Reile, Oberstleutnant, Abwehr III, 23
Resistance (France), sabotage, 79
Reyburn, Wallace, 217; *Rehearsal for Invasion*, 197
Ribbentrop, Joachim von, 86
Roberts, Major-General J.H., 1, 26
Robertson, Colonel T.A., 49, 50, 59
Robertson, Terence, *Dieppe, The Shame and the Glory*, 18, 25

Robinson, J.R., 'Radar Intelligence and the Dieppe Raid', 204-5
Rommel, Field Marshal Erwin, 182, 221
Roskill, Captain Stephen, 18, 24; and Irving, 13-17, 22; *The War at Sea*, 15, 16
Roundup (Operation), 64, 211, 212-13
Royal Air Force *see* RAF
Royal Navy, 7, 134, 150-1, 223-4
Royal Scotsman (LSI), 97
RSHA (Amt VI), German intelligence organization, 42, 49
Ruge, Vice-Admiral F.: Naval Defences West, 93, 104, 107-8, 112; coastal radar chain, 125-7
Rundstedt, Field Marshal von, 43, 79, 82, 84-5, 102; and agents' reports, 45, 55; Battle Report, 28-9, 59, 96, 180-2; reorganization of defences, 200, 214
Rusbridger, James, 160
Russia, pressure for Second Front from, 63, 68
Russian front: German preoccupation with, 80, 82, 84, 89, 90; Sigint about German operations on, 183
Rutter (Operation): cancellation of, 4-5, 51, 150, 151; and movement of 10 Pz division, 25-6, 27; planning of, 26-7, 69, 95, 212-13; plans leaked as deception, 67; and radar, 124, 135-7; and strategic surprise, 149

Saalwächter, General Admiral, Naval Group West, 112
sabotage in France, 79
St Cecily, Plage, radar station, 128, 129
St Jean de Luz, 27, 98
St Malo, 105

St Nazaire (Operation Chariot), 17, 65, 67, 173, 213; documents captured at, 175; Hitler affronted by, 82; radar and, 127-8, 133; role of Ultra at, 161; surprise achieved, 98
St Nazaire (Sledgehammer West), 67
Salerno, 210
Sartorius, Oberstleutnant i G, 133, 141
Scharnhorst (battleship), 173
Schnösenberg, Captain, 143
Schwabenland (aircraft catapult ship), 130
Scotland, Expeditionary Force in, 91, 92
scrambler telephones, 171
Sea Commanders (*Seekommandanten*), 133-4, 145
Sea Lion (Operation), 94, 96, 103, 105, 106
Second Front: German expectation of, 22, 80-1, 83, 84-7; public demand for, 63, 67-8, 69-70
Second Front, *see also* invasion
Secret Intelligence Service (SIS/MI 6), 25
security: for Jubilee, 5-6, 57-8; on south coast, 4-5
Seeadler (German torpedo boat), 104, 126
Seine, Bay of the, 45, 90, 92-3, 94, 102, 112
SGB (steam gun boats), 11, 104
shipping: British shortage of, 64, 87, 89; German attacks on, 79, 90; German lack of, 96, 104; increase in small craft, 78; radar equipped and controlled (British), 104
Sigint (signal intelligence), 24, 42, 180, 182, 183, 217, *see also* Ultra
Simkins, C.A.G. (with Hinsley), *Security and Counter-Intelligence*, 27-8
Slazak (destroyer), 109, 148

INDEX

Sledgehammer Central and West (Operations), 67
Sledgehammer East (Operation), 26, 67, 152, 213
Sledgehammer (Operation), 57, 63–5, 185
Small Ships Code, 176–7
smoke, use of, 12, 98, 217
Solo One (Operation), 46
Southampton, 61, 78
Special Intelligence (SI), 159, 167, 170–1, 173–4, *see also* Ultra
'special means', in deception plans, 52, 57–8, 70
Speer, Reichsminister, 80
Stacey, Colonel C.P., 25, 28, 213; Official History of the Canadian Army, 2, 16
Stalin, Joseph, 89
Stanley, Colonel Oliver, 51–2, 62, 63
Starkey (Operation), 55, 207, 208, 209
Stier (Schiff 23), 103, 104
strategic deception, 27, 198
strategic surprise, 149, 153
Sucharov, Major B., Royal Canadian Engineers, 222
Suez, importance of, 88
surprise *see* strategic surprise; tactical surprise
Swanage, radar station at, 139
Swinton, Lord, 149
Sylvan Flakes soap advertisement, 21, 52

T-boats (Torpedo boats), 103, 104, 106, 175
tactical doctrine for assaults, 215–16
tactical intelligence, Hinsley on, 25
tactical surprise: importance of, 25, 28, 69, 70, 149–50, 152; in Neptune planning, 225
Tait, Air Vice-Marshal Sir Victor, 202, 205
Tank Design, Department of, 222
tanks: AVRE, 222; CDL (Canal Defence Light), 222, 224; Churchill, 9–10; heavy assault, 218; role of, 222; Sherman medium, 222; specialized, 222, 225; waterproofed Churchill, 218
tanks, *see also* landing craft, LCT (Tanks)
tides, and timings, 94, 95, 110
timetables, for Jubilee, 7–8
timing: day and night, 7–8, 219–20, 223, 225; tides and, 94, 95, 110
Tizard, Sir Henry, 125
Todt Organization, 50, 103
Torch (Operation), 42, 89, 216; cover plans for, 46, 48, 62, 65; and Jubilee, 199
training, effect of uncertainty on, 63
Trevor-Roper, H.R., *Hitler's War Directives, 1939–1945*, 17
Turquoise (motorship), 104, 135, 176
Twenty Committee *see* agents; XX (Twenty) Committee

U-boats, 80, 89, 90, 103, 113; vulnerability of bases, 85, 86, 91
Ulm (German minelayer), sinking of, 170
Ulster Monarch (LSI), 97, 188
Ultra: delays in decrypts, 167, 171, 172; and effect of radar counter- measures, 203; extreme security of, 169–71, 174, 175, 176; and GAF, 161, 188–9; and German defence measures, 161–2, 221; influence of on operations, 30–1, 163–4, 167–72; intelligence about enemy dispositions, 172–4; and Jubilee, 30, 161, 166, 167, 169, 185; release of decrypts, 31, 159, 165; revealed by Winterbotham, 24, 159
Ultra, *see also* Enigma; Sigint
Unwin, Major, warnings about caves, 151
US forces, 43, 64–5
US National Archives, material in, 32

US Navy, detail of planning foreign to, 220
USA, and concept of invasion plan, 63, 64–5, 226

Ventnor, radar station at, 147, 149
Villa, Brian Loring, *Unauthorized Action: Mountbatten and the Dieppe Raid*, 2, 33–4

W Committee, to co-ordinate deception, 27
Wagner, Captain, 94
Warlimont, Lieutenant-General Walter, 46, 84, 90
Watt, Professor D.C., 31
Wavell, General, deception planning, 62
weather, significance of, 4, 110, 151, 152
Weber, Oberleutnant, 29, 133, 139, 141, 143
Weg Rosa, German shipping channel, 103, 104
Weigley, Russell, *Eisenhower's Lieutenants*, 220
Wells, H.G., 70
Wernher, Sir Harold, 226
West, Nigel, *Unreliable Witness*, 48, 49
Wetbob (Operation), 140–1
Wheatley, Flight Lieutenant Dennis, 62
Whitworth Jones, Air Commodore, 187
Wight, Isle of, 5, 45, 62, 79, 149
Williams, Francis, 206

Winterbotham, Group Captain F.W., *The Ultra Secret*, 24, 30, 159
wire, German defensive, 214
wireless procedure (British), changed before Jubilee, 180, 181–2
wireless transmission: assumption that B-Dienst would read, 176–7; German, 204; and intelligence of German mine-laying, 174–5
wireless transmission, *see also* Y intelligence
Worthing, minefields at, 60, 61
Wurmbach, Oberleutnant z S, 123
Wyburd, Commander, (5 Group), 148

XX (Twenty) Committee, 22, 51, 70, *see also* agents, double
XXIV Maggio (ore ship), 112

Y intelligence: GAF, 178–9, 182; Oculist, 138, 139, 145–6, 203, 205; primacy of for German High Command, 43; RAF, 178
Y intelligence, *see also* wireless transmission
Yarmouth, 79
Yukon exercises, 182, 224; radar and, 132, 139
Yukon I exercise, 4, 70, 79, 151
Yukon II exercise, 4, 70, 79, 102, 178

Zeitzler, Major-General K., 80, 85–6
Ziegler, Philip, *Mountbatten*, Jubilee, 2, 34, and Ultra, 164–5

For Product Safety Concerns and Information please contact our EU representative GPSR@taylorandfrancis.com
Taylor & Francis Verlag GmbH, Kaufingerstraße 24, 80331 München, Germany

www.ingramcontent.com/pod-product-compliance
Lightning Source LLC
Chambersburg PA
CBHW070559300426
44113CB00010B/1313